Compiling Algorithms
for Heterogeneous Systems

Synthesis Lectures on Computer Architecture

Editor
Margaret Martonosi, *Princeton University*

Founding Editor Emeritus
Mark D. Hill, *University of Wisconsin, Madison*

Synthesis Lectures on Computer Architecture publishes 50- to 100-page publications on topics pertaining to the science and art of designing, analyzing, selecting and interconnecting hardware components to create computers that meet functional, performance and cost goals. The scope will largely follow the purview of premier computer architecture conferences, such as ISCA, HPCA, MICRO, and ASPLOS.

Compiling Algorithms for Heterogeneous Systems

Steven Bell, Jing Pu, James Hegarty, and Mark Horowitz

ISBN: 978-3-031-00630-2   paperback
ISBN: 978-3-031-01758-2   ebook
ISBN: 978-3-031-00055-3   hardcover

DOI 10.1007/978-3-031-01758-2

A Publication in the Springer Nature series
SYNTHESIS LECTURES ON COMPUTER ARCHITECTURE

Lecture #43
Series Editor: Margaret Martonosi, *Princeton University*
Founding Editor Emeritus: Mark D. Hill, *University of Wisconsin, Madison*
Series ISSN
Print 1935-3235   Electronic 1935-3243

Compiling Algorithms
for Heterogeneous Systems

Steven Bell
Stanford University

Jing Pu
Google

James Hegarty
Oculus

Mark Horowitz
Stanford University

SYNTHESIS LECTURES ON COMPUTER ARCHITECTURE #43

ABSTRACT

Most emerging applications in imaging and machine learning must perform immense amounts of computation while holding to strict limits on energy and power. To meet these goals, architects are building increasingly specialized compute engines tailored for these specific tasks. The resulting computer systems are heterogeneous, containing multiple processing cores with wildly different execution models. Unfortunately, the cost of producing this specialized hardware—and the software to control it—is astronomical. Moreover, the task of porting algorithms to these heterogeneous machines typically requires that the algorithm be partitioned across the machine and rewritten for each specific architecture, which is time consuming and prone to error.

Over the last several years, the authors have approached this problem using domain-specific languages (DSLs): high-level programming languages customized for specific domains, such as database manipulation, machine learning, or image processing. By giving up generality, these languages are able to provide high-level abstractions to the developer while producing high-performance output. The purpose of this book is to spur the adoption and the creation of domain-specific languages, especially for the task of creating hardware designs.

In the first chapter, a short historical journey explains the forces driving computer architecture today. Chapter 2 describes the various methods for producing designs for accelerators, outlining the push for more abstraction and the tools that enable designers to work at a higher conceptual level. From there, Chapter 3 provides a brief introduction to image processing algorithms and hardware design patterns for implementing them. Chapters 4 and 5 describe and compare Darkroom and Halide, two domain-specific languages created for image processing that produce high-performance designs for both FPGAs and CPUs from the same source code, enabling rapid design cycles and quick porting of algorithms. The final section describes how the DSL approach also simplifies the problem of interfacing between application code and the accelerator by generating the driver stack in addition to the accelerator configuration.

This book should serve as a useful introduction to domain-specialized computing for computer architecture students and as a primer on domain-specific languages and image processing hardware for those with more experience in the field.

KEYWORDS

domain-specific languages, high-level synthesis, compilers, image processing accelerators, stencil computation

Contents

Preface

Cameras are ubiquitous, and computers are increasingly being used to process image data to produce better images, recognize objects, build representations of the physical world, and extract salient bits from massive streams of video, among countless other things. But while the data deluge continues to increase, and while the number of transistors that can be cost-effectively placed on a silicon die is still going up (for now), limitations on power and energy mean that traditional CPUs alone are insufficient to meet the demand. As a result, architects are building more and more specialized compute engines tailored to provide energy and performance gains on these specific tasks.

Unfortunately, the cost of producing this specialized hardware—and the software to control it—is astronomical. Moreover, the resulting computer systems are heterogeneous, containing multiple processing cores with wildly different execution models. The task of porting algorithms to these heterogeneous machines typically requires that the algorithm be partitioned across the machine and rewritten for each specific architecture, which is time consuming and prone to error.

Over the last several years, we have approached this problem using domain-specific languages (DSLs)—high-level programming languages customized for specific domains, such as database manipulation, machine learning, or image processing. By giving up generality, these languages are able to provide high-level abstractions to the developer while producing high-performance output. Our purpose in writing this book is to spur the adoption and the creation of domain-specific languages, especially for the task of creating hardware designs.

This book is not an exhaustive description of image processing accelerators, nor of domain-specific languages. Instead, we aim to show why DSLs make sense in light of the current state of computer architecture and development tools, and to illustrate with some specific examples what advantages DSLs provide, and what tradeoffs must be made when designing them. Our examples will come from image processing, and our primary targets are mixed CPU/FPGA systems, but the underlying techniques and principles apply to other domains and platforms as well. We assume only passing familiarity with image processing, and focus our discussion on the architecture and compiler sides of the problem.

In the first chapter, we take a short historical journey to explain the forces driving computer architecture today. Chapter 2 describes the various methods for producing designs for accelerators, outlining the push for more abstraction and the tools that enable designers to work at a higher conceptual level. In Chapter 3, we provide a brief introduction to image processing algorithms and hardware design patterns for implementing them, which we use through the rest of the book. Chapters 4 and 5 describe Darkroom and Halide, two domain-specific languages created for image processing. Both are able to produce high-performance designs for both FPGAs and CPUs from the same source code, enabling rapid design cycles and quick porting of algorithms. We present both of these examples because comparing and contrasting them illustrates some of the tradeoffs and design decisions encountered when creating a DSL. The final portion of the book discusses the task of controlling specialized hardware within a heterogeneous system running a multiuser operating system. We give a brief overview of how this works on Linux and show how DSLs enable us to automatically generate the necessary driver and interface code, greatly simplifying the creation of that interface.

This book assumes at least some background in computer architecture, such as an advanced undergraduate or early graduate course in CPU architecture. We also build on ideas from compilers, programming languages, FPGA synthesis, and operating systems, but the book should be accessible to those without extensive study on these topics.

Steven Bell, Jing Pu, James Hegarty, and Mark Horowitz
January 2018

Acknowledgments

Any work of this size is necessarily the result of many collaborations. We are grateful to John Brunhaver, Zachary DeVito, Pat Hanrahan, Jonathan Ragan-Kelley, Steve Richardson, Jeff Setter, Artem Vasilyev, and Xuan Yang, who influenced our thinking on these topics and helped develop portions of the systems described in this book. We're also thankful to Mike Morgan, Margaret Martonosi, and the team at Morgan & Claypool for shepherding us through the writing and production process, and to the reviewers whose feedback made this a much better manuscript than it would have been otherwise.

Steven Bell, Jing Pu, James Hegarty, and Mark Horowitz
January 2018

CHAPTER 1

Introduction

When the International Technology Roadmap for Semiconductors organization announced its final roadmap in 2016, it was widely heralded as the official end of Moore's law [ITRS, 2016]. As we write this, 7 nm technology is still projected to provide cheaper transistors than current technology, so it isn't over just yet. But after decades of transistor scaling, the ITRS report revealed at least modest agreement across the industry that cost-effective scaling to 5 nm and below was hardly a guarantee.

While the death of Moore's law remains a topic of debate, there isn't any debate that the nature and benefit of scaling has decreased dramatically. Since the early 2000s, scaling has not brought the power reductions it used to provide. As a result, computing devices are limited by the electrical power they can dissipate, and this limitation has forced designers to find more energy-efficient computing structures. In the 2000s this power limitation led to the rise of multicore processing, and is the reason that practically all current computing devices (outside of embedded systems) contain multiple CPUs on each die. But multiprocessing was not enough to continue to scale performance, and specialized processors were also added to systems to make them more energy efficient. GPUs were added for graphics and data-parallel floating point operations, specialized image and video processors were added to handle video, and digital signal processors were added to handle the processing required for wireless communication.

On one hand, this shift in structure has made computation more energy efficient; on the other, it has made programming the resulting systems much more complex. The vast majority of algorithms and programming languages were created for an abstract computing machine running a single thread of control, with access to the entire memory of the machine. Changing these algorithms and languages to leverage multiple threads is difficult, and mapping them to use the specialized processors is near impossible. As a result, accelerators only get used when performance is essential to the application; otherwise, the code is written for CPU and declared "good enough." Unless we develop new languages and tools that dramatically simplify the task of mapping algorithms onto these modern heterogeneous machines, computing performance will stagnate.

This book describes one approach to address this issue. By restricting the application domain, it is possible to create programming languages and compilers that can ease the burden of creating and mapping applications to specialized computing resources, allowing us to run complete applications on heterogeneous platforms. We will illustrate this with examples from image processing and computer vision, but the underlying principles extend to other domains.

The rest of this chapter explains the constraints that any solution to this problem must work within. The next section briefly reviews how computers were initially able to take advantage of Moore's law scaling without changing the programming model, why that is no longer the case, and why energy efficiency is now key to performance scaling. Section 1.2 then shows how to compare different power-constrained designs to determine which is best. Since performance and power are tightly coupled, they both need to be considered to make the best decision. Using these metrics, and some information about the energy and area cost of different operations, this section also points out the types of algorithms that benefit the most from specialized compute engines. While these metrics show the potential of specialization, Section 1.3 describes the costs of this approach, which historically required large teams to design the customized hardware and develop the software that ran on it. The remaining chapters in this book describe one approach that addresses these cost issues.

1.1   CMOS SCALING AND THE RISE OF SPECIALIZATION

From the earliest days of electronic computers, improvements in physical technology have continually driven computer performance. The first few technology changes were discrete jumps, first from vacuum tubes to bipolar transistors in the 1950s, and then from discrete transistors to bipolar integrated circuits (ICs) in the 1960s. Once computers were built with ICs, they were able to take advantage of Moore's law, the prediction-turned-industry-roadmap which stated that the number of components that could be economically packed onto an integrated circuit would double every two years [Moore, 1965].

As MOS transistor technology matured, gates built with MOS transistors used less power and area than gates built with bipolar transistors, and it became clear in the late 1970s that MOS technology would dominate. During this time Robert Dennard at IBM Research published his paper on MOS scaling rules, which showed different approaches that could be taken to scale MOS transistors [Dennard et al., 1974]. In particular, he observed that if a transistor's operating voltage and doping concentration were scaled along with its physical dimensions, then a number of other properties scaled nicely as well, and the resized transistor would behave predictably.

If a MOS transistor is shrunk by a factor of $1/\kappa$ in each linear dimension, and the operating voltage is lowered by the same $1/\kappa$, then several things follow:

1. Transistors get smaller, allowing κ^2 more logic gates in the same silicon area.

2. Voltages and currents inside the transistor scale by a factor of $1/\kappa$.

3. The effective resistance of the transistor, I/V, remains constant, due to 2 above.

4. The gate capacitance C shrinks by a factor of $1/\kappa$ ($1/\kappa^2$ due to decreased area, multiplied by κ due to reduced electrode spacing).

The switching time for a logic gate is proportional to the resistance of the driving transistor multiplied by the capacitance of the driven transistor. If the effective resistance remains constant while the capacitance decreases by $1/\kappa$, then the overall delay also decreases by $1/\kappa$, and the chip can be run faster by a factor of κ.

Taken together, these scaling factors mean that κ^2 more logic gates are switched κ faster, for a total increase of κ^3 more gate evaluations per second. At the same time, the energy required to switch a logic gate is proportional to CV^2. With both capacitance and voltage decreasing by a factor of $1/\kappa$, the energy per gate evaluation decreased by a factor of $1/\kappa^3$.

During this period, roughly every other year, a new technology process yielded transistors which were about $1/\sqrt{2}$ as large in each dimension. Following Dennard scaling, this would give a chip with twice as many gates and a faster clock by a factor of 1.4×, making it 2.8× more powerful than the previous one. Simultaneously, however, the energy dissipated by each gate evaluation dropped by 2.8×, meaning that total power required was the same as the previous chip. This remarkable result allowed each new generation to achieve nearly a 3× improvement for the same die area and power.

This scaling is great in theory, but what happened in practice is somewhat more circuitous. First, until the mid-1980s, most complex ICs were made with nMOS rather than CMOS gates, which dissipate power even when they aren't switching (known as static power). Second, during this period power supply voltages remained at 5 V, a standard set in the bipolar IC days. As a result of both of these, the power per gate did not change much even as transistors scaled down. As nMOS chips grew more complex, the power dissipation of these chips became a serious problem. This eventually forced the entire industry to transition from nMOS to CMOS technology, despite the additional manufacturing complexity and lower intrinsic gate speed of CMOS.

After transitioning to CMOS ICs in the mid-1980s, power supply voltages began to scale down, but not exactly in sync with technology. While transistor density and clock speed continued to scale, the energy per logic gate dropped more slowly. With the number of gate evaluations per second increasing faster than the energy of gate evaluation was scaling down, the overall chip power grew exponentially.

This power scaling is exactly what we see when we look at historical data from CMOS microprocessors, shown in Figure 1.1. From 1980 to 2000, the number of transistors on a chip increased by about 500× (Figure 1.1a), which corresponds to scaling transistor feature size by roughly 20×. During this same period of time, processor clock frequency increased by 100×, which is 5× faster than one would expect from simple gate speed (Figure 1.1b). Most of this additional clock speed gain came from microarchitectural changes to create more deeply pipelined "short tick" machines with fewer gates per cycle, which were enabled by better circuit designs of key functional units. While these fast clocks were good for performance, they were bad from a power perspective.

By 2000, computers were executing 50,000× more gate evaluations per second than they had in the 1980s. During this time the average capacitance had scaled down, providing a 20× energy savings, but power supply voltages had only scaled by 4–5× (Figure 1.1c), giving roughly a 25× savings. Taken together the capacitance and supply scaling only reduce the gate energy by around 500×, which means that the power dissipation of the processors should increase by two orders of magnitude during this period. Figure 1.1d shows that is exactly what happened.

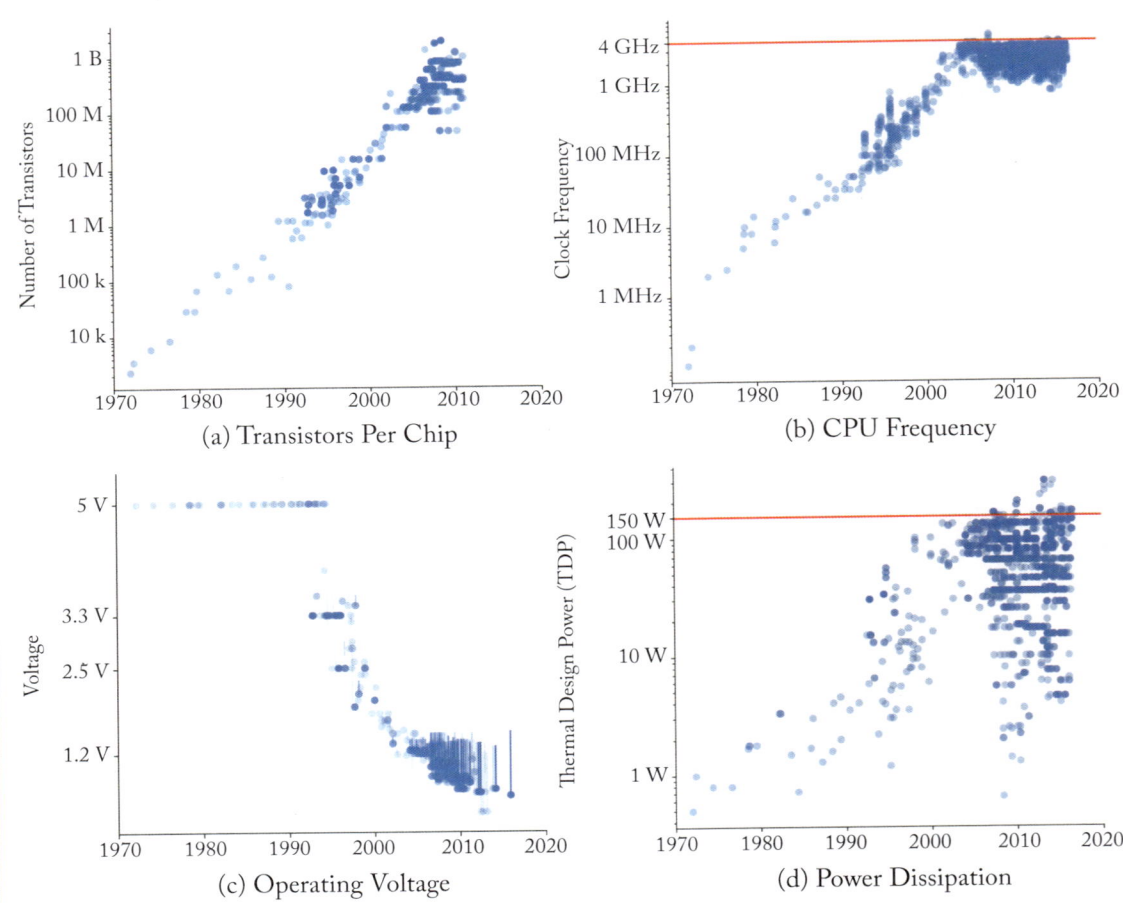

Figure 1.1: From the 1960s until the early 2000s, transistor density and operating frequency scaled up exponentially, providing exponential performance improvements. Power dissipation increased but was kept in check by lowering the operating voltage. Data from CPUDB [Danowitz et al., 2012].

Up to this point, all of these additional transistors were used for a host of architectural improvements that increased performance even further, including pipelined datapaths, superscalar instruction issue, and out-of-order execution. However, the instruction set architectures (ISAs) for various processors generally remained the same through multiple hardware revisions, meaning that existing software could run on the newer machine without modification—and reap a performance improvement.

But around 2004, Dennard scaling broke down. Lowering the gate threshold voltage further caused the leakage power to rise unacceptably high, so it began to level out just below 1 V.

Without the possibility to manage the power density by scaling voltage, manufacturers hit the "power wall" (the red line in Figure 1.1d). Chips such as the Intel Pentium 4 were dissipating a little over 100 W at peak performance, which is roughly the limit of a traditional package with a heatsink-and-fan cooling system. Running a CPU at significantly higher power than this requires an increasingly complex cooling system, both at a system level and within the chip itself.

Pushed up against the power wall, the only choice was to stop increasing the clock frequency and find other ways to increase performance. Although Intel had predicted processor clock rates over 10 GHz, actual numbers peaked around 4 GHz and settled back between 2 and 4 GHz (Figure 1.1b).

Even though Dennard scaling had stopped, taking down frequency scaling with it, Moore's law continued its steady march forward. This left architects with an abundance of transistors, but the traditional microarchitectural approaches to improving performance had been mostly mined out. As a result, computer architecture has turned in several new directions to improve performance without increasing power consumption.

The first major tack was symmetric multicore, which stamped down two (and then four, and then eight) copies of the CPU on each chip. This has the obvious benefit of delivering more computational power for the same clock rate. Doubling the core count still doubles the total power, but if the clock frequency is dialed back, the chip runs at a lower voltage, keeping the energy constant while maintaining some of the performance advantage of having multiple cores. This is especially true if the parallel cores are simplified and designed for energy efficiency rather than single-thread performance. Nonetheless, even simple CPU cores incur significant overhead to compute their results, and there is a limit to how much efficiency can be achieved simply by making more copies.

The next theme was to build processors to exploit regularity in certain applications, leading to the rise of single-instruction-multiple-data (SIMD) instruction sets and general-purpose GPU computing (GPGPU). These go further than symmetric multicore in that they amortize the instruction fetch and decode steps across many hardware units, taking advantage of data parallelism. Neither SIMD nor GPUs were new; SIMD had existed for decades as a staple of supercomputer architectures and made its way into desktop processors for multimedia applications along with GPUs in the late 1990s. But in the mid-2000s, they started to became prominent as a way to accelerate traditional compute-intensive applications.

A third major tack in architecture was the proliferation of specialized accelerators, which go even further in stripping out control flow and optimizing data movement for particular applications. This trend was hastened by the widespread migration to mobile devices and "the cloud," where power is paramount and typical use is dominated by a handful of tasks. A modern smartphone System-on-chip (SoC) contains more than a dozen custom compute engines, created specifically to perform intensive tasks that would be impossible to run in real time on the main CPU. For example, communicating over WiFi and cellular networks requires complex coding and modulation/demodulation, which is performed on a small collection of hardware units specialized for these signal processing tasks. Likewise, decoding or encoding video—whether for watching Netflix, video chatting, or camera filming—is handled by hardware blocks that only perform this specific task. And the process of capturing raw pixels and turning them into a pleasing (or at least presentable) image is performed by a long pipeline of hardware units that demosaic, color balance, denoise, sharpen, and gamma-correct the image.

Even low-intensity tasks are getting accelerators. For example, playing music from an MP3 file requires relatively little computational work, but the CPU must wake up a few dozen times per second to fill a buffer with sound samples. For power efficiency, it may be better to have a dedicated chip (or accelerator within the SoC, decoupled from the CPU) that just handles audio.

While there remain some performance gains still to be squeezed out of thread and data parallelism by incrementally advancing CPU and GPU architectures, they cannot close the gap to a fully customized ASIC. The reason, as we've already hinted, comes down to power.

Cell phones are power-limited both by their battery capacity (roughly 8–12 Wh) and the amount of heat it is acceptable to dissipate in the user's hand (around 2 W). The datacenter is the same story at a different scale. A warehouse-sized datacenter consumes tens of megawatts, requiring a dedicated substation and a cheap source of electrical power. And like phones, data center performance is constrained partly by the limits of our ability to get heat out, as evidenced by recent experiments and plans to build datacenters in caves or in frigid parts of the ocean. Thus, in today's power-constrained computing environment, the formula for improvement is simple: performance per watt is performance.

Only specialized architectures can optimize the data storage and movement to achieve the energy reduction we want. As we will discuss in Section 1.4, specialized accelerators are able to eliminate the overhead of instructions by "baking" them into the computation hardware itself. They also eliminate waste for data movement by designing the storage to match the algorithm.

Of course, general-purpose processors are still necessary for most code, and so modern systems are increasingly heterogeneous. As mentioned earlier, SoCs for mobile devices contain dozens of processors and specialized hardware units, and datacenters are increasingly adding GPUs, FPGAs, and ASIC accelerators [AWS, 2017, Norman P. Jouppi et al., 2017].

In the remainder of this chapter, we'll describe the metrics that characterize a "good" accelerator and explain how these factors will determine the kind of systems we will build in the future. Then we lay out the challenges to specialization and describe the kinds of applications for which we can expect accelerators to be most effective.

1.2   WHAT WILL WE BUILD NOW?

Given that specialized accelerators are—and will continue to be—an important part of computer architecture for the foreseeable future, the question arises: What makes a good accelerator? Or said another way, if I have a potential set of designs, how do I choose what to add to my SoC or datacenter, if anything?

1.2.1   PERFORMANCE, POWER, AND AREA

On the surface, the good things we want are obvious. We want high performance, low power, and low cost.

Raw performance—the speed at which a device is able to perform a computation—is the most obvious measure of "good-ness." Consumers will throw down cash for faster devices, whether that performance means quicker web page loads or richer graphics. Unfortunately, this isn't easy to quantify with the most commonly advertised metrics.

Clock speed matters, but we also need to account for how much work is done on each clock cycle. Multiplying clock speed by the number of instructions issued per cycle is better, but still ignores the fact that some instructions might do much more work than others. And on top of this, we have the fact that utilization is rarely 100% and depends heavily on the architecture and application.

We can quantify performance in a device-independent way by counting the number of essential operations performed per unit time. For the purposes of this metric, we define "essential operations" to include only the operations that form the actual result of the computation. Most devices require a great deal of non-essential computation, such as decoding instructions or loading and storing intermediate data. These are "non-essential" not because they are pointless or unnecessary but because they are not intrinsically required to perform the computation. They are simply overhead incurred by the specific architecture.

With this definition, adding two pieces of data to produce an intermediate result is an essential operation, but incrementing a loop counter is not since the latter is required by the implementation and not the computation itself.

To make things concrete, a 3×3 convolution on a single-channel image requires nine multiplications (multiplying 3×3 pixels by their corresponding weights) and eight 2-input additions per output pixel. For a 640×480 image (307,200 pixels), this is a little more than 5.2 million total operations.

A CPU implementation requires many more instructions than this to compute the result since the instruction stream includes conditional branches, loop index computations, and so forth. On the flip side, some implementations might require fewer instructions than operations, if they process multiple pieces of data on each instruction or have complex instructions that fuse multiple operations. But implementations across this whole spectrum can be compared if we calculate everything in terms of device-independent operations, rather than device-specific instructions.

The second metric is power consumption, measured in Watts. In a datacenter context, the power consumption is directly related to the operating cost, and thus to the total cost of ownership (TCO). In a mobile device, power consumption determines how long the battery will last (or how large a battery is necessary for the device to survive all day). Power consumption also determines the maximum computational load that can be sustained without causing the device to overheat and throttle back.

The third metric is cost. We'll discuss development costs further in the following section, but for now it is sufficient to observe that the production cost of the final product is closely related to the silicon area of the chip, typically measured in square millimeters (mm^2). More chips of a smaller design will fit on a fixed-size wafer, and smaller chips are likely to have somewhat higher yield percentages, both of which reduce the manufacturing cost.

However, as important as performance, power, and silicon area are as metrics, they can't be used directly to compare designs, because it is relatively straightforward to trade one for the other.

Running a chip at a higher operating voltage causes its transistors to switch more rapidly, allowing us to increase the clock frequency and get increased performance, at the cost of increased power consumption. Conversely, lowering the operating voltage along with the clock frequency saves energy, at the cost of lower performance.[1]

It isn't fair to compare the raw performance of a desktop Intel Core i7 to an ARM phone SoC, if for no other reason than that the desktop processor has a 20–50× power advantage. Instead, it is more appropriate to divide the power (Joules per second) by the performance (operations per second) to get the average energy used per computation (Joules per operation). Throughout the rest of this book, we'll refer to this as "energy per operation" or pJ/op. We could equivalently think about maximizing the inverse: operations/Joule.

For a battery-powered device, energy per operation relates directly to the amount of computation that can be performed with a single battery charge; for anything plugged into the wall, this relates the amount of useful computation that was done with the money you paid to the electric company.

A similar difficulty is related to the area metric. For applications with sufficient parallelism, we can double performance simply by stamping down two copies of the same processor on a chip. This benefit requires no increase in clock speed or operating voltage—only more silicon. This was, of course, the basic impetus behind going to multi core computation.

Even further, it is possible to lower the voltage and clock frequency of the two cores, trading performance for energy efficiency as described earlier. As a result, it is possible to improve either power or performance by increasing silicon area as long as there is enough parallelism. Thus, when comparing between architectures for highly parallel applications, it is helpful to

---

[1]Of course, modern CPUs do this scaling on the fly to match their performance to the ever-changing CPU load, known as "Dynamic Voltage and Frequency Scaling" (DVFS).

normalize performance by the silicon area used. This gives us operations/Joule divided by area, or $\frac{\text{ops}}{\text{mm}^2 \cdot \text{J}}$.

These two compound metrics, p.J/operation and $\frac{\text{ops}}{\text{mm}^2 \cdot \text{J}}$, give us meaningful ways to compare and evaluate vastly different architectures. However, it isn't sufficient to simply minimize these in the abstract; we must consider the overall system and application workload.

1.2.2   FLEXIBILITY

Engineers building a system are concerned with a particular application, or perhaps a collection of applications, and the metrics discussed are only helpful insofar as they represent performance on the applications of interest. If a specialized hardware module cannot run our problem, its energy and area efficiency are irrelevant. Likewise, if a module can only accelerate parts of the application, or only some applications out of a larger suite, then its benefit is capped by Ahmdahl's law. As a result, we have a flexibility tradeoff: more flexible devices allow us to accelerate computation that would otherwise remain on the CPU, but increased flexibility often means reduced efficiency.

Suppose a hypothetical fixed-function device can accelerate 50% of a computation by a factor of $100\times$, reducing the total computation time from 1 second to 0.505 seconds. If adding some flexibility to the device drops the performance to only $10\times$ but allows us to accelerate 70% of the computation, we will now complete the computation in 0.37 seconds—a clear win.

Moreover, many applications demand flexibility, whether the product is a networking device that needs to support new protocols or an augmented-reality headset that must incorporate the latest advances in computer vision. As more and more devices are connected to the internet, consumers increasingly expect that features can be upgraded and bugs can be fixed via over-the-air updates. In this market, a fixed-function device that cannot support rapid iteration during prototyping and cannot be reconfigured once deployed is a major liability.

The tradeoff is that flexibility isn't free, as we have already alluded to. It almost always hurts efficiency (performance per watt or $\frac{\text{ops}}{\text{mm}^2 \cdot \text{J}}$) since overhead is spent processing the configuration. Figure 1.2 illustrates this by comparing the performance and efficiency for a range of designs proposed at ISSCC a number of years ago. While newer semiconductor processes have reduced energy across the board, the same trend holds: the most flexible devices (CPUs) are the least efficient, and increasing specialization also increases performance, by as much as three orders of magnitude.

In certain domains, this tension has created something of a paradox: applications that were traditionally performed completely in hardware are moving toward software implementations, even while competing forces push related applications away from software toward hardware. For example, the fundamental premise of software defined radio (SDR) is that moving much (or all) of the signal processing for a radio from hardware to software makes it possible to build a system that is simpler, cheaper, and more flexible. With only a minimal analog front-end, an SDR system can easily run numerous different coding and demodulation schemes, and be upgraded

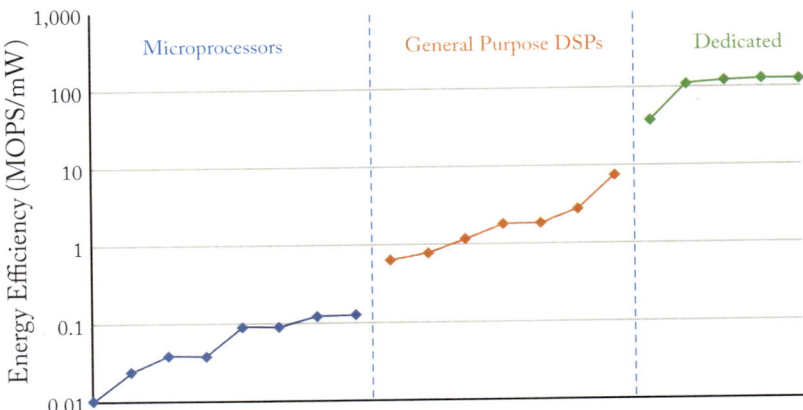

Figure 1.2: Comparison of efficiency for a number of designs from ISSCC, showing the clear tradeoff between flexibility and efficiency. Designs are sorted by efficiency and grouped by overall design paradigm. Figure from Marković and Brodersen [2012].

over the air. But because real-time signal processing requires extremely high computation rates, many SDR platforms use an FPGA, and carefully optimized libraries have been written to fully exploit the SIMD and digital signal processing (DSP) hardware in common SoCs. Likewise, software-defined networking aims to provide software-based reconfigurability to networks, but at the same time more and more effort is being poured into custom networking chips.

1.3   THE COST OF SPECIALIZATION

To fit these metrics together, we must consider one more factor: cost. After all, given the enormous benefits of specialization, the only thing preventing us from making a specialized accelerator for everything is the expense.

Figure 1.3 compares the non-recurring engineering (NRE) cost of building a new high-end SoC on the past few silicon process nodes. The price tags for the most recent technologies are now well out of reach for all but the largest companies. Most ASICs are less expensive than this, by virtue of being less complex, using purchased or existing IP, having lower performance targets, and being produced on older and mature processes [Khazraee et al., 2017]. Yet these costs still run into the millions of dollars and remain risky undertakings for many businesses.

Several components contribute to this cost. The most obvious is the price of the lithography masks and tooling setup, which has been driven up by the increasingly high precision of each process node. Likewise, these processes have ever-more-stringent design rules, which require more engineering effort during the place and route process and in verification. The exponential increase in number of transistors has enabled a corresponding growth in design complexity, which comes with increased development expense. Some of these additional transistors are used

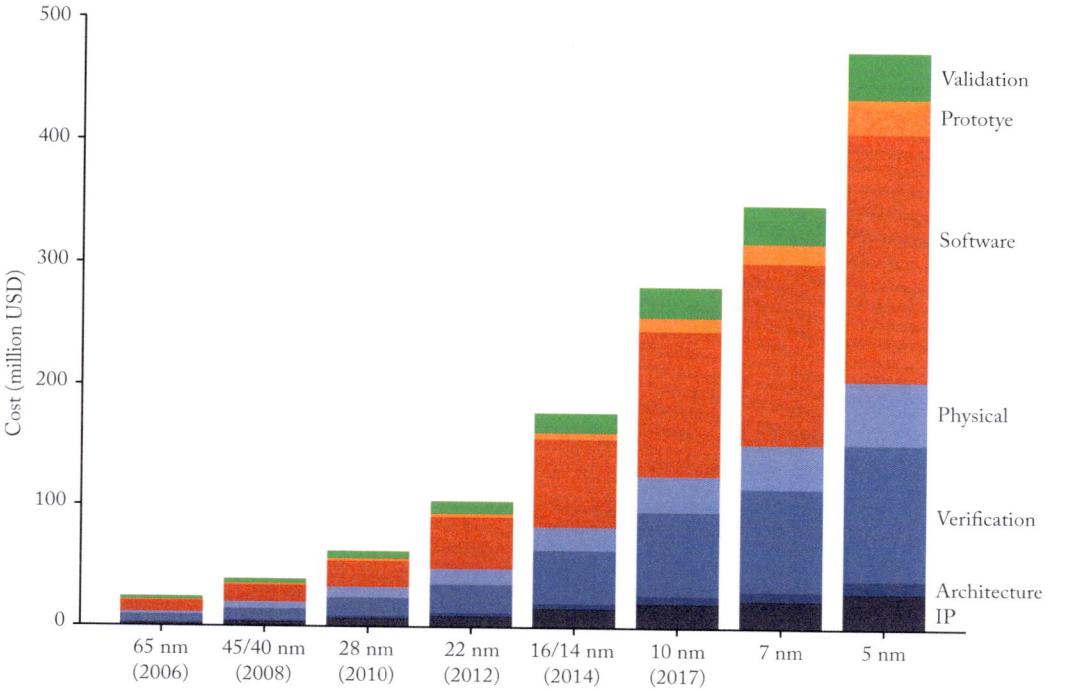

Figure 1.3: Estimated cost breakdown to build a large SoC. The overall cost is increasing exponentially, and software comprises nearly half of the total cost. (Data from International Business Strategies [IBS, 2017].)

in ways that do not appreciably increase the design complexity, such as additional copies of processor cores or larger caches. But while the exact slope of the correlation is debatable, the trend is clear: More transistors means more complexity, and therefore higher design costs. Moreover, with increased complexity comes increased costs for testing and verification.

Last, but particularly relevant to this book, is the cost of developing software to run the chip, which in the IBS estimates accounts for roughly 40% of the total cost. The accelerator must be configured, whether with microcode, a set of registers, or something else, and it must be interfaced with the software running on the rest of the system. Even the most rigid of "fixed" devices usually have some degree of configurability, such as the ability to set an operating mode or to control specific parameters or coefficients.

This by itself is unremarkable, except that all of these "configurations" are tied to a programming model very different than the idealized CPU that most developers are used to. Timing details become crucial, instructions execute out of order or in a massively parallel fashion, and concurrency and synchronization are handled with device-specific primitives. Accelerators are, almost by definition, difficult to program.

To state the obvious, the more configurable a device is, the more effort must go into configuring it. In highly configurable accelerators such as GPUs or FPGAs, it is quite easy—even typical—to produce configurations that do not perform well. Entire job descriptions revolve around being able to work the magic to create high-performance configurations for accelerators. These people, informally known as "the FPGA wizards" or "GPU gurus," have an intimate knowledge of the device hardware and carry a large toolbox of techniques for optimizing applications. They also have excellent job security.

This difficulty is exacerbated by a lack of tools. Specialized accelerators need specialized tools, often including a compiler toolchain, debugger, and perhaps even an operating system. This is not a problem in the CPU space: there are only a handful of competitive CPU architectures, and many groups are developing tools, both commercial and open source. Intel is but one of many groups with an x86 C++ compiler, and the same is true for ARM. But specialized accelerators are not as widespread, and making tools for them is less profitable. Unsurprisingly, NVIDIA remains the primary source of compilers, debuggers, and development tools for their GPUs. This software design effort cannot easily be pushed onto third-party companies or the open-source community, and becomes part of the chip development cost.

As we stand today, bringing a new piece of silicon to market is as much about writing software as it is designing logic. It isn't sufficient to just "write a driver" for the hardware; what is needed is an effective bridge to application-level code.

Ultimately, companies will only create and use accelerators if the improvement justifies the expense. That is, an accelerator is only worthwhile if the engineering cost can be recouped by savings in the operating cost, or if the accelerator enables an application that was previously impossible. The operating cost is closely tied to the efficiency of the computing system, both in terms of the number of units necessary (buying a dozen CPUs vs. a single customized accelerator) and in terms of time and electricity. Because it is almost always easier to implement an algorithm on a more flexible device, this cost optimization results in a tug-of-war between performance and flexibility, illustrated in Figure 1.4.

This is particularly true for low-volume products, where the NRE cost dominates the overall expense. In such cases, the cheapest solution—rather than the most efficient—might be the best. Often, the most cost-effective solution to speed up an application is to buy a more powerful computer (or a whole rack of computers!) and run the same horribly inefficient code on it. This is why an enormous amount of code, even deployed production code, is written in languages like Python and MATLAB, which have poor runtime performance but terrific developer productivity.

Our goal is to reduce the cost of developing accelerators and of mapping emerging applications onto heterogeneous systems, pushing down the NRE of the high-cost/high-performance

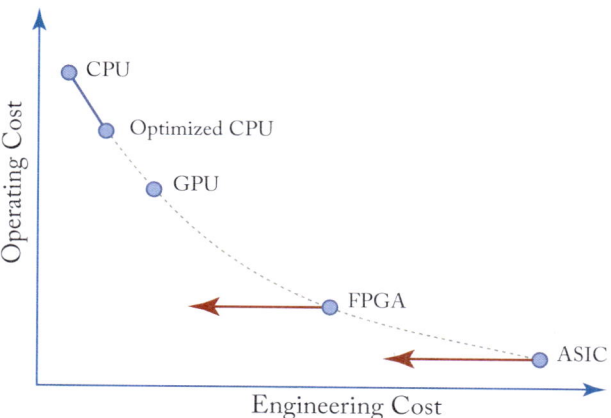

Figure 1.4: Tradeoff of operating cost (which is inversely related to runtime performance) vs. non-recurring engineering cost (which is inversely related to flexibility). More flexible devices (CPUs and GPUs) require less development effort but achieve worse performance compared to FPGAs and ASICs. We aim to reduce the engineering development cost (red arrows), making it more feasible to adopt specialized computing.

areas of this tradeoff space. Unless we do so, it will remain more cost effective to use general-purpose systems, and computer performance in many areas will suffer.

1.4   GOOD APPLICATIONS FOR ACCELERATION

Before we launch into systems for programming accelerators, we'll examine which applications can be accelerated most effectively. Can all applications be accelerated with specialized processors, or just some of them?

The short answer is that only a few types of applications are worth accelerating. To see why, we have to go back to the fundamentals of power and energy. Given that, for a modern chip, performance per watt is equivalent to performance, we want to minimize the energy consumed per unit of computation. That is, if the way to maximize operations per second is to maximize operations per second per watt, we can cancel "seconds," and simply maximize operations per Joule.

Table 1.1 shows the energy required for a handful of fundamental operations in a 45 nm process. The numbers are smaller for more recent process nodes, but the relative scale remains essentially the same.

The crucial observation here is that a DRAM fetch requires 500× more energy than a 32-bit multiplication, and 50,000× more than an 8-bit addition. The cost of fetching data from memory completely dwarfs the cost of computing with it. The cache hierarchy helps, of course,

Table 1.1: Approximate energy required for various data operations in a 45 nm process at 0.9 V. Data from Brunhaver [2015].

Operation	Energy (pJ)
8-bit add	0.03
8-bit multiply	0.2
32-bit add	0.1
32-bit multiply	3.1
32-bit floating-point add	0.9
32-bit floating-point multiply	3.7
8 kB cache fetch	10
32 kB cache fetch	20
1 MB cache fetch	100
DRAM fetch	1,300–2,600

but it can't completely solve the problem. If the data must be refetched for every computation, then the overall performance will be limited by the memory system, and the actual compute hardware matters very little. Said another way, the only way for a computation to be extremely efficient is to do as much computation as possible for each data fetch.

Thus, our best candidates for acceleration are applications with tons of computation per memory fetch. This implies two things: first, that the application performs a ton of computation, and second, that it does that computation on data it has already read into memory and cached nearby. In other words, the computation must be large and have very high data locality.

What applications have this type of extreme data locality and high compute to data fetch ratio? A few candidates include:

- signal processing, with convolution and other filters running on dense arrays of samples;

- image and video processing, which is much like signal processing but often with even more data (typically with two or perhaps three spatial dimensions);

- computer vision, particularly neural networks and especially convolutional neural networks (CNNs); and

- dense linear algebra (e.g., solvers for large optimization problems).

To be sure, this is a limited class of applications. However, it is a particularly important class of applications, and these are all growing rapidly in importance. Mobile image processing and computational photography are becoming commonplace on phones, with burst photography and HDR imaging running by default in Android and iOS. However, the computational intensity of these algorithms means they are limited to still images.

Emerging applications such as augmented reality need very low latency processing for lots of pixels, in a form factor that can fit into a headset or even glasses. Displays are moving toward 4K video, and new mobile phones are capable of capturing 1080p video at over 100 frames per second, so the demand to process more pixels is not about to abate.

In machine learning, neural networks have gone from an academic curiosity to champion algorithm precisely because of access to massive computational resources. Over the last few years, we have seen predictions improve with larger networks and more data, which suggests that this demand will only increase.

This is not to say that other applications—such as graph processing or database workloads—cannot be accelerated. But unless the data can be reorganized for high locality, these will be ultimately limited by the memory.

For the remainder of this book, we will focus on image processing with stencil pipelines, which we'll describe in detail in Chapter 3. Although this is a relatively focused domain, the concepts presented here are applicable to other domains with similarly high locality and computation density.

In the following chapter, we will discuss how configurations for various accelerators are created, and what compilers do (or should do) to make that task easier, lowering the costs of creating a design. We follow with two examples of domain-specific languages that map image processing applications into high-performance implementations on both CPUs and FPGAs, showing how domain-specific languages help solve the dilemmas faced by traditional compilers and synthesis systems. Finally, we turn our focus to integration in heterogeneous systems and explore how these systems are put together from a system and software perspective.

CHAPTER 2

Computations and Compilers

We have argued that we must reduce the cost of developing and programming specialized accelerators if silicon computing is to continue improving at the rates we are accustomed to. The question is, how?

Throughout the history of computing, we have harnessed the complexity of our systems in two ways: first, by adding layers of abstraction, and second, by creating tools that are able to automate large parts of the design process. The key benefit of abstraction is that the developer can work at a conceptual level much higher than, and completely independent of, the physical machine where the design is implemented. And as abstraction increases, so does the need for a tool that can elaborate upon the abstraction to produce a machine-level instantiation.

Ultimately, reducing development cost will hinge on our ability to create more powerful tools: tools that provide natural high-level abstractions, and which can automate more of the design process to produce high-performing designs. In this chapter, we discuss several approaches to developing designs for heterogeneous systems, with a special view toward mixed CPU/FPGA-based accelerators for image processing. We'll explain how differing "machine models" and levels of abstraction can ease or aggravate this task, and point toward the development of the new tools that we discuss in the following chapters.

Computing systems come in a dizzying variety of forms, from completely general-purpose CPUs to semi-specialized processors like GPUs and DSPs, to reconfigurable computing fabrics like FPGAs and coarse-grained reconfigurable arrays (CGRAs) to custom chips for a single design (ASICs). Despite the diversity, however, all computing devices necessarily have some notion of a "program," which defines what computation they perform. To have any hope of mapping an algorithm onto a heterogeneous system, we must first consider what constitutes a "program" on various architectures, and what is similar and different as we translate that program across systems.

All computing platforms expose a set of fundamental building blocks, which are used to compose the computation. We'll call these *design primitives*, although they take different names in different systems. In a traditional CPU, the fundamental blocks are the instructions, defined by the instruction set architecture (ISA). The computation is defined by a sequence of instructions, which are executed by the processor. On an FPGA, the building blocks are typically look up tables (LUTs), memory blocks, and specialized arithmetic units. The computation is defined not by a temporal or sequential order but by the bits that control the LUT outputs and define the interconnections between components.

Likewise, the computation of an ASIC is defined by the physical layout of logic gates, and the wiring that connects them together on the chip. We could push on to lower levels of composition—transistors and wires and doping regions—but practically speaking, most digital designers work at or above the gate level, and it is here that practically all digital computing systems share a common basis. Whatever the underlying mechanisms, nearly every system can perform bitwise Boolean operations, arithmetic, comparisons, and some type of conditional operation.

A machine model is the interface that a particular device or programming language presents to a developer. At the lowest level, the machine model is the physical hardware interface: registers, instructions, and so forth. At higher levels, detailed conceptual models also function as machine models, independent of the actual hardware. For example, an out-of-order processor presents an in-order machine model to the software, and this becomes the foundation for thinking about and modeling the computation.

2.1   DIRECT SPECIFICATION

We can specify a design for a device by using its fundamental design primitives, working directly with the machine model that the hardware implements. The obvious example of direct specification is writing a CPU program in machine assembly. An assembler provides some important conveniences (such as calculating the address of a jump), but the machine model in the developer's head is the same as the one that will execute the code. The assembler simply translates human-readable machine instructions into machine-readable machine instructions.

But very little CPU code is written this way, just as very few FPGA and ASIC designs are created by directly writing netlists of look up tables and Boolean logic gates. The reason is simple: modern applications are simply too large and complex to be developed at this level. Another less obvious drawback is the fact that with direct specification, *what* is computed becomes inextricably mixed with *how* it is computed. This has two consequences. First, it is nearly impossible to "port" the design to another system, because the algorithm only exists as a set of machine-specific design primitives. Second, optimization choices are blended together with the algorithm, and it is impossible to modify one without rethinking the other. As a result, the developer must have a strong knowledge of the underlying hardware, and it is nearly impossible to avoid premature optimization. When your building blocks are fine-grained and close to the hardware, it is hard not to optimize them, even if the real performance problems are at a higher level.

But before we dismiss direct specification as too difficult and time consuming, it's important to point out that it has advantages, too. When done well, direct specification permits extremely high performance. The developer has complete control over the hardware, and is able to take full advantage of its resources. It is not uncommon for a few bits of a crucial CPU function to be written in assembly in order to squeeze out the last drop of performance in a subroutine where it will make a difference. Likewise, high-performance ASIC designs rely on a small number of hand-optimized blocks. But because of the prohibitive cost, only a tiny fraction of a design is created this way. Everything else builds on these optimized sub-units, or remains unoptimized.

2.2 COMPILERS

Because working directly with the machine primitives is extremely burdensome, countless programming languages have been created that allow developers to work with higher-level abstractions.

As a first step, consider the C programming language. In many ways, C is a thin veneer over the ISA of a typical processor: its mathematical and logical operations map directly to assembly instructions, and its implementations of pointers and control flow don't go much further. But C (along with practically all other languages) has some key features beyond nicer syntax for expressing mathematical operations, and these have made it incredibly successful.

Rather than work with a fixed number of registers (as a real CPU must), C envisions an abstract machine with infinite registers (or perhaps equivalently, without registers at all), where all data is stored in a flat memory space. The developer need not be concerned with which variable is stored where, and this burden is instead placed on the compiler. C also adds support for function calls with arbitrary arguments, again by having the compiler do the underlying work of building stack frames and copying the appropriate arguments. Finally, while C was designed for CPUs, and while it defines a program much like assembly (as sequences of instructions that operate on values), C does not describe a series of operations on any particular processor. As a result, it is at least somewhat architecture-independent, and a C compiler exists for almost every CPU-like architecture that has ever been created.

In a similar way, the Verilog and VHDL hardware design languages introduce constructs for clocked logic, allowing developers to work with a simplified conceptual model of digital design.

These languages work because of compilers: automatic tools that take a design built on a set of design patterns and produce a translation into the fundamental primitives for a specific system. Said again, a compiler takes a program specification for an abstract machine model and fills in the details to produce a specification for a real machine.

We'll call this process *lowering*, since the design patterns of the source are "higher level" (more coarse-grained) than the design patterns of the target. They lack the detail of a lower-level implementation, and that detail is supplied by the compiler. In general, lowering is straightforward. It is mostly a matter of finding the most effective implementation for a particular construct and instantiating that everywhere it appears.

Today, most designs are created with compilers for a number of good reasons. Compilers enable much higher productivity than direct specification, and well-designed languages facilitate software composition so that it is at least feasible to manage the complexity of a large design.

Software libraries, hardware IP catalogs, and package managers all make it easier to reuse parts of a design, further accelerating development.

But CPU languages like C and hardware languages like Verilog and VHDL also suffer many of the same drawbacks as direct specification. First, despite the major productivity gains versus direct specification, systems created with C and Verilog still require enormous design effort.

Even though C is portable across many CPU architectures, high-performance C code is not. To achieve high performance, a developer ends up coding for the CPU—adding device-specific calls for vectorization, blocking or tiling the data for cache locality, and allocating work across multiple threads. These optimizations can yield a 10× performance improvement or better [Kelley et al., 2013], but because the optimizations are specific to the hardware, the resulting code is no longer portable. In the same way, although Verilog or VHDL code can be mapped to practically any programmable logic device, high performance often requires that the developer apply device-specific optimizations, such as using special clock buffers or specifying logic placement constraints.

To mitigate these drawbacks, we have a continuum of higher-level languages: Java code compiles to the abstract "Java Virtual Machine," which was created to improve portability across CPU architectures. Java also adds features like garbage collection, which—together with its memory allocation mechanisms—raises the level of abstraction for memory. Heap memory is automatically allocated and cleaned up by the system, (mostly) removing this burden from the developer. Dynamically typed languages such as Python raise the level even higher. Memory is handled entirely by the runtime system, and the developer works primarily with abstract data structures like lists and tuples.

Higher level languages leave more details unspecified in the user code, which means the developer has less work to do and the compiler more. In many cases, especially where developer time is at a premium and code performance is unimportant, this is a win. But no compiler is perfect, and the more lowering a compiler has to do, the harder it is to produce high-performance code. For example, a generic list structure is likely to have a single implementation that is optimized for an "average" use case. Particular use cases might be be amenable to far more efficient implementations, but the generality of the language dictates that everything be implemented that way. For example, a fixed-length array of integers doesn't need the flexibility of a Python list, which can contain other objects, be appended to, and so forth. The fact that Python doesn't allow the developer to manage memory and doesn't permit pointer access to arrays of data means that some operations are stupendously inefficient. In practice, these operations are implemented in another language and linked in (e.g., the NumPy numerical computation library, or Pillow, the successor to the Python Imaging Library).

To summarize, raising the level of abstraction with a higher-level language is extremely beneficial for both portability and productivity, but it often comes with a serious performance penalty. Two questions arise as we extend these ideas to hardware: First, given that increased abstraction offers increased portability, how much abstraction do we need for code to be portable between CPUs, FPGAs, and possibly other accelerators in a heterogeneous system? And, second, if we start with that level of abstraction, what does it take to produce a high-performance implementation?

2.3   HIGH-LEVEL SYNTHESIS

We now turn specifically to hardware designs for ASIC and FPGA. Unlike CPUs, where there are dozens of programming languages with widespread adoption, hardware design is dominated by Verilog and VHDL, which are both relatively low-level.

Given that many developers already know C/C++, and that these represent a level of abstraction above Verilog and VHDL, it seems appealing to try to compile C code to register transfer language (RTL) code. Although C is a low-level language as far as CPU code is concerned, it is "high level" with respect to hardware RTL in that it does not specify any details about architecture or timing. This approach has been termed *high-level synthesis* (HLS), and it aims to improve productivity by decoupling the design description from low-level issues like register allocation, clock-level timing, and pipelining.

An HLS compilation process usually can be divided into four major steps: program analysis, allocation, scheduling, and binding [Coussy et al., 2009]. First, the compiler statically analyzes the input program to extract the data dependency of operations and the control flow that drives it. The strength of the analysis and the optimization that follows strongly determine the quality of the result, and these are usually limited by the level of abstraction and the style of the input source code. It is relatively straightforward to apply automatic pipelining for an untimed C function that consists of single assignments to variables. A generic C program is a different matter, with variable reuse and complex control flow making it difficult (or impossible) to analyze the dataflow.

Once analysis is complete, allocation determines the type and the number of hardware units (e.g., functional units, memory elements) required to implement the compute operations, controls, and storage of the program. It also pulls the characteristics of the hardware units (such as timing, area, and power) from a target platform specific database to be used in the remaining HLS tasks. Next, scheduling assigns all operations in the program to specific clock cycle slots and creates a finite state machine that controls the execution of these operations in sequence. Finally, binding does the final assignment of operations to physical hardware units and the allocation of registers on the clock boundaries. In the binding phase, compilers often seek opportunities to share hardware resources among operations that are mutually exclusive (e.g., inputs to a multiplexer) or have non-overlapping lifetimes.

In addition to providing a higher level abstraction for hardware development, HLS can produce highly optimized implementations for certain classes of problems, including image processing. As input designs are untimed, by using information about the target platform, HLS tools can easily schedule a large graph of arithmetic operations into a hardware pipeline with an optimal number of stages to achieve a target clock frequency [Zhang et al., 2008]. Furthermore, since the input is C/C++ code (or perhaps another CPU language), it is possible to compile and run the input code on a CPU for functional simulation. The C-based functional simulations of HLS designs are much faster than cycle-accurate simulation in HDL, enabling rapid development cycles.

HLS has been built into many commercial EDA tools, including Vivado HLS [Xilinx, 2016], Catapult HLS [Mentor Graphics, 2016], Altera OpenCL [Altera, 2016] and MaxCompiler [Technology, 2011] and has gained a wide adoption in industry and academia. As a result, HLS tools are seeing increasing use and increasing research interest, especially for FPGAs [Cattaneo et al., 2015, Chen et al., 2016, Moreau et al., 2015, Zhang et al., 2015].

Despite the advantages, HLS suffers from the issues of productivity and portability. HLS is far more efficient than writing RTL for some tasks, but contrary to the claims of the marketing literature, writing good HLS code still requires a solid grasp of hardware architecture since these systems can't yet consume generic code and partition it optimally for locality and parallelism. Furthermore, HLS developers have the additional burden of directly managing local memories in order to exploit locality, which is similar to the scenario of programming a scratchpad-based processor. In the absence of hardware-managed caches, programmers not only need to organize the order of the computation to fit the working sets in the device memory hierarchy, but must also allocate different levels of local buffers manually and derive data transfers among them.

Portability remains an equally serious problem. High-performance HLS code uses an entirely different set of design patterns than high-performance CPU code, and so an optimized HLS implementation sometimes looks completely different from the code of the same application targeted for traditional processors. As a consequence, porting and optimizing an application often requires global restructuring of its code—which is time consuming and prone to error. And as with C/C++ on a CPU, functionality and optimization are conflated in the application code, making it hard to change one without affecting the other.

As an example, Figure 2.1 shows pseudo-code for CPU and FPGA implementations of a line buffer feeding a 3×3 convolution stage. Though both pieces of the code implement line buffers (i.e., three lines of input pixels) and avoid recomputation of input pixels, the FPGA code is more complicated as it includes explicit management of the input stencil buffer. In Figure 2.1a, each iteration of the loop over x (4) computes one output pixel and new input pixels, while the input values already in the line buffer are skipped by using conditional expressions for the lower bounds of the loops in 6 and 7. Note that the CPU implementation assumes that the data reuse of adjacent input stencils will be automatically exploited by caches and register files of processors as long as the reuse distance of data is minimized by the loop structure in the code. In contrast, in Figure 2.1b, the FPGA implementation has to explicitly allocate a 3×3 stencil buffer (3) and manage the data movement within the stencil buffer (13 to 16) and between the stencil buffer and the line buffer (17 to 19).

```
1  alloc output [8, 8]   // Input is 10x10, output is 8x8
2  // For every OUTPUT pixel
3  for y = 0 to 7
4    for x = 0 to 7
5      // Compute enough input pixels, but skip values that are already computed
6      for win.y = (y==0 ? 0 : 2) to 2
7        for win.x = (x==0 ? 0 : 2) to 2
8          compute input at [x+win.x, y+win.y], store in input[x+win.x, (y+win.y)%3]
9      // Compute an output pixel by convolving 3x3 filter weights
10     for win.y = 0 to 2
11       for win.x = 0 to 2
12         output[x, y] += input[x+win.x, (y+win.y)%3] * weight[win.x, win.y]
13
```

(a) CPU Target

```
1  alloc input [10, 3]
2  alloc output [8, 8]
3  alloc stencil [3, 3]    // 3x3 input stencil
4  // For every INPUT pixel
5  for y = 0 to 9
6    for x = 0 to 9
7      // Up shift a column of input buffer to evict an old value
8      for row = 0 to 1
9        input[x, row] = input[x, row - 1]
10     // Read in input and fill the buffer
11     compute input at [x, y], store it in input[x, 2]
12     if y >= 2  // Not on top border
13       // Left shift values in stencil
14       for win.y = 0 to 2
15         for win.x = 0 to 1
16           stencil[win.x, win.y] = stencil[win.x + 1, win.y]
17       // Read a column from input buffer
18       for row = 0 to 2
19         stencil[2, row] = input[x + 2, row]
20       if x >= 2  // Not on left border
21         // Compute an output pixel by convolving input stencil and weights
22         for win.y = 0 to 2
23           for win.x = 0 to 2
24             output[x-2, y-2] += stencil[win.x, win.y] * weight[win.x, win.y]
```

(b) FPGA Target

Figure 2.1: Pseudo-code of 3×3 convolution with a line buffer for CPU and FPGA targets.

HLS tools provide a more effective way to create hardware designs for certain classes of problems than traditional RTL languages, but because they must "raise" code from a CPU-oriented language before lowering it to RTL, HLS code must be carefully crafted for the compiler. This is shown visually in Figure 2.2, which lays out the space of languages we have discussed so far.

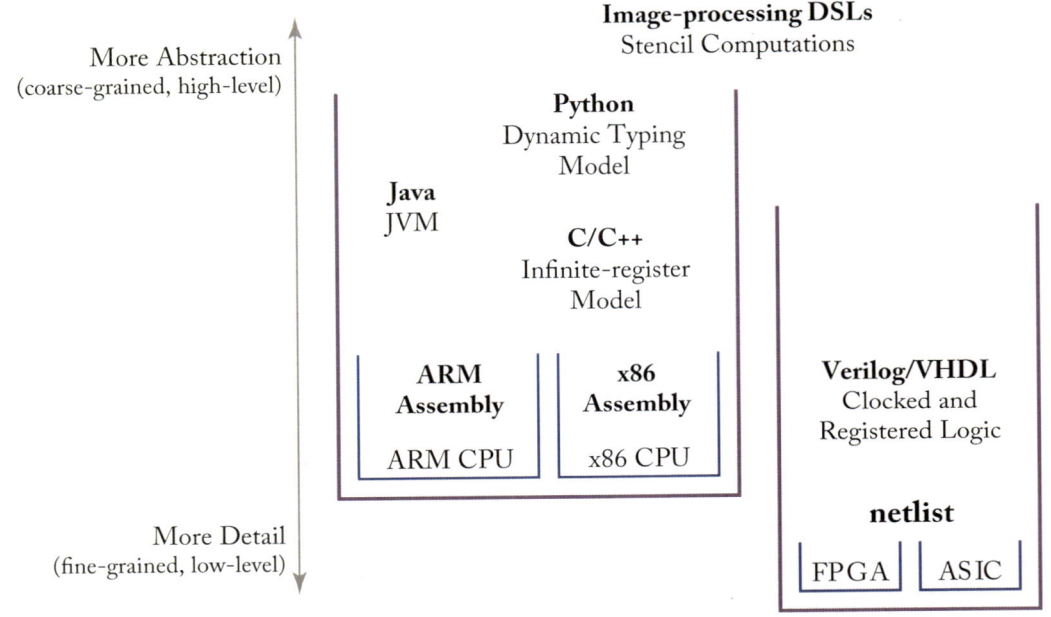

Figure 2.2: Visual landscape of the languages discussed in this book. The vertical axis represents abstraction, with higher-level languages at the top. Each silo represents the space of languages that targets each particular machine model.

2.4 DOMAIN-SPECIFIC LANGUAGES

So what if, rather than starting with C/C++ code and stripping out the CPU-oriented control flow to derive the computation, we simply started with that computation? HLS has to do some "raising" before it can do any lowering; what if we had a programming language that would allow us to specify the algorithm at that higher level that could then be lowered into RTL? That is precisely the promise of domain-specific languages (DSLs).

Domain-specific languages are not a radically new concept; instead, they arise as a natural interface for many software abstractions. In one sense, DSLs are just libraries or tools with useful abstractions whose interface constitutes a language. For example, the MATLAB language began as an interactive command-line wrapper around LINPACK, a library for matrix computation. Similarly, Unix tools like `sed` and `awk` provide a high-level interface for text processing. Although `sed` and `awk` are typically used for one-line commands at a command prompt, the interfaces they expose are in themselves languages. Perl began in a very similar way to `sed` and `awk` but was soon adopted for broader purposes as a general scripting language. Likewise, the operation of GNU Make is controlled by a `Makefile`. In one sense, a `Makefile` is just a complex configuration for a tool, but in another sense, the `Makefile` command structure is a language—perhaps limited and specialized, but a language nonetheless. These are only a few examples, but thousands of these "little languages" have been invented for various special purposes [Bentley, 1986]. They work because they provide convenient and powerful control over abstractions for a particular class of problems.

As a result, general-purpose programming languages are offering more and better tools that enable developers to build these kinds of abstraction interfaces. A ubiquitous example is operator overloading, where the semantics of basic operators are given additional meanings. A C++ linear algebra library can overload operators for *, /, and +, providing a natural interface to the underlying library calls. This becomes a small language of its own, albeit embedded in C++.

Macros, template metaprogramming, and Python's "decorators" go even further, providing constructs for code to generate and even modify other bits of code. These language facilities can be easily leveraged to build tools that parse, elaborate, optimize, and generate code. Such tools remain embedded in their host languages, but they are, in essence, compilers.

Said again, DSLs are libraries or tools that provide domain-specific abstractions in the form of a language. Following the trajectory of this chapter, DSLs raise the level of abstraction yet again, leaving even more implementation details unspecified. This simplifies the job of the developer while providing the freedom to lower the code into an implementation for many different devices. However, this also comes with a cost: it won't do to simply wave our hands and raise the level of abstraction; the implementation details must come from somewhere.

It's precisely at this point that domain specialization becomes valuable. As with HLS, there are a relatively small number of design patterns that have effective mappings to hardware. The same is true of CPUs, GPUs, and other accelerators. For a given type of problem, there are only a few reasonable implementations. By making the language specific to a particular domain, we narrow down the set of possible implementations that a compiler needs to choose from, which greatly simplifies the lowering task. Put bluntly, the trick to building a high-performance DSL is to take out all the features that cannot be executed efficiently.

SQL is effective because it is a *Structured* Query Language. The language defines a limited set of directives for making queries, and database backends are highly optimized to handle these operations. Similarly, graphics APIs like OpenGL and DirectShow offer a restricted API primarily designed to render triangles, because triangles can be rendered very efficiently. Other shapes, such as ellipses or splines, are rendered simply by approximating them with triangles and straight line segments and rendering those. Graphic shading languages operate on an equally restricted model. A vertex shader cannot access neighboring vertices, despite the fact that such access would enable all sorts of interesting algorithms. Pixel shaders are likewise unable to operate on neighboring pixels. These kinds of restrictions limit what can be expressed, but allow the rendering pipeline to be mapped onto the massively parallel hardware of a GPU.

The advantage, as we will show in the next few chapters, is that the DSL approach allows developers to create high-performance systems with very little design effort compared to other approaches. If designed well, DSL code is highly portable and can be retargeted for multiple architectures with very few changes to the code. The obvious disadvantage is that a domain-specific language can only express algorithms within a restricted domain. However, as we argued in Section 1.4, the set of applications that can be accelerated to great effect is limited, so a handful of DSLs will be able to cover the most important areas.

After a brief discussion of image processing, the following chapters will take a close look at two particular DSLs and discuss how they facilitate this process of creating high-performance implementations from high-productivity and portable code.

CHAPTER 3

Image Processing with Stencil Pipelines

In this chapter, we'll examine image processing in more depth, showing why it is particularly amenable to hardware acceleration, and describing the architectural patterns that we will leverage in the remainder of the book.

Image processing is an especially important domain for acceleration because it requires immense amounts of computation. A 1080p/30fps video stream contains $1920 \times 1080 \times 30 \approx$ 62 million pixels per second. Even a very simple filter that requires only 30 operations per pixel would therefore require nearly 2 billion operations/second, enough to keep a CPU very busy.

Image processing is also fertile for acceleration because the majority of image operations are performed on small neighborhoods of pixels, allowing algorithms to have very high locality. For example, two-dimensional convolution only requires a neighborhood as large as the convolution kernel. Likewise, dilation/erosion, peak detection, and other operations also function on small neighborhoods of nearby pixels. These can be computed in a "sliding window" fashion where the neighborhoods required to compute adjacent output pixels overlap substantially. With careful management, pixel values can be cached locally and reused rather than reloaded from main memory. In the best case, it is possible to access the input pixels only once, meaning that the image can be "streamed" through the accelerator.

3.1 IMAGE SIGNAL PROCESSORS

An excellent example of an image-processing accelerator is a camera "image signal processor" (ISP), which transforms the raw pixels coming off the sensor into a presentable image. Nearly all digital cameras have an ISP, whether integrated into the sensor, as part of a mobile SoC, or as a standalone chip. They perform several billion pixel operations per second, which would be impossible on a CPU, and they do it on a power budget that allows cameras to record video for hours off a battery.

In a conventional ISP, raw pixels are converted into an image through a long sequence of operations, usually including:

- **black-level correction**, which subtracts the non-zero offset value that pixels measure due to dark current (a function of temperature and other factors);

- **dead-pixel correction**, which fills in data for known bad pixels, by interpolating from the surrounding values;

- **demosaicking**, which interpolates RGB values from the Bayer mosaic. In most color image sensors, each pixel only senses one of red, green, and blue, and the demosaicking step fills in the other two colors for each pixel;

- **lens shading correction**, which corrects for differences in illumination strength across the image sensor. Depending on the lens, pixels at the corners can receive less than half the light that center pixels receive; and

- **white balance** and **color correction**, which attempt to transform the measured colors (as a function of the current lighting and sensor color sensitivity characteristics) into a color space suitable for storage and display.

Some of these operations, such as black-level correction, operate on only a single pixel at a time. Others, such as demosaicking, read from a neighborhood of pixels around the point of interest, which we'll call a *stencil*. The pixels within the stencil are combined with some set of operations, which we'll call a *stencil function*, as shown in Figure 3.1.

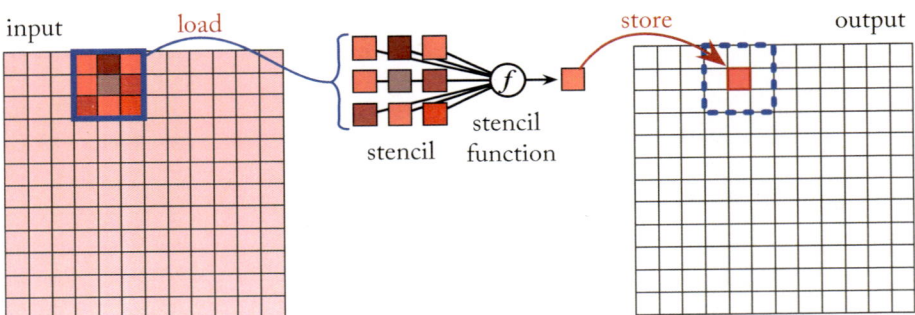

Figure 3.1: Computation of an output image with a stencil function. Neighborhoods of pixels from the input image ("stencils") are fed into a stencil function, which computes a single output pixel.

In an ISP, the raw input pixels arrive from the image sensor line by line, in row-major order. As each stage computes its result, the intermediate values are stored in a buffer for consumption by the next stage. Because each computation stage only takes values from within the fixed-size stencil, the intermediate buffer does not need to store the whole image; only a few rows are needed. More precisely, for an n-high $\times$ m-wide stencil, the buffer needs to store $n - 1$ rows plus m pixels, illustrated in Figure 3.2. Because this buffer stores lines of the image, it is (not surprisingly) referred to as a *line buffer*. The ISP is a good example of what we will refer to as a "line-buffered pipeline": a series of stencil operations connected by line buffers.

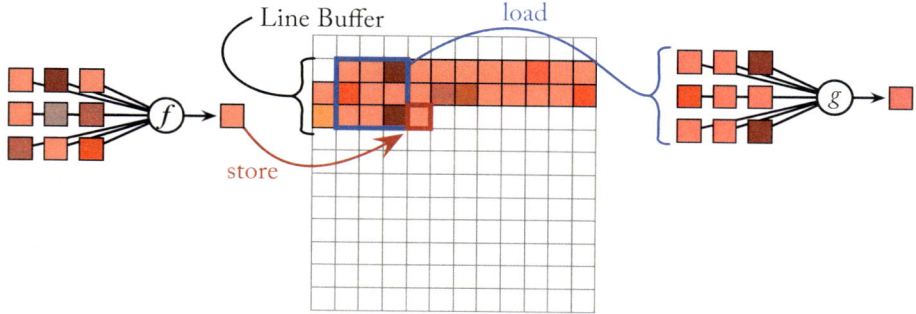

Figure 3.2: Chaining of stencil computations together with a line buffer. For a 3×3 stencil, only two lines of the image plus three pixels need to be stored, provided that pixels are produced and consumed at the same rate.

3.2   EXAMPLE APPLICATIONS

Line-buffered pipelines are good for ISPs, but the same structure is also effective for a wide range of image-processing applications. Here we briefly describe a number of applications that serve as helpful examples, and which we will use to evaluate the systems we describe later.

Campipe is an implementation of a camera ISP. It includes basic raw conversion operations (demosaicking, white balance, and color correction), plus enhancement and error correction operations (crosstalk correction, dead pixel suppression, and black level correction).

Convolution is a two-dimensional convolution, which can represent a variety of image processing filters, including sharpening and blurring. Where convolution is used to implement a Gaussian blur, we'll refer to it as **gaussian**.

Unsharp sharpens an image by calculating a blurred copy of the image, subtracting this from the original image to get the sharp edges (high frequencies), and then adds some portion of this "edge image" back to the original to obtain a sharpened output.

Corner is the classic Harris corner detection algorithm [Harris and Stephens, 1988], which is used as an early stage in many computer vision algorithms to pick out point features for tracking. The Harris algorithm begins by taking the vertical and horizontal gradient I_x and I_y at each pixel and forming a 2×2 matrix:

$$\begin{bmatrix} I_x^2 & I_x I_y \\ I_x I_y & I_y^2 \end{bmatrix}$$

It then finds an approximation for the eigenvalues of this matrix and selects points where both eigenvalues are high. In essence, this finds pixels that have strong orthogonal edges.

Edge is the Canny edge detection algorithm [Canny, 1986]. It finds the magnitude of the gradient of the image in x and y, sums them, and then thins the edges using non-maximum

Figure 3.3: Some example applications that are implemented in later sections of this book.

suppression. Pixels with a sufficiently high gradient are labeled "edge pixels." Finally, pixels adjacent to edge pixels are also marked as edge pixels if their gradient is above a second, lower threshold. This final step is computed recursively, tracing edges as far as they go.

Deblur is an implementation of the Richardson-Lucy non-blind deconvolution algorithm [Richardson, 1972]. It is an iterative algorithm, which alternately attempts deconvolution with an estimated blur kernel and then uses the deconvolved image to produce a better estimate of the blur.

Flow implements the Lucas-Kanade algorithm for dense optical flow [Lucas et al., 1981]. Optical flow serves as input to many higher-level computer vision algorithms. It is often implemented as a multi scale algorithm, which uses an image pyramid to efficiently search a large area [Bouguet, 2001].

Stereo computes depth from a stereo image pair. First, it rectifies the left and right images based on known camera calibration parameters so that all correspondences are on the same horizontal line in the image. Then, for each pixel in the left channel, it searches 80 horizontal pixels neighboring that point in the right channel, evaluating a 9×9 sum of absolute differences (SAD) between the two images. The correspondence with the lowest SAD gives the most likely depth. STEREO is a simple pipeline, but it performs an enormous amount of computation due to the large search window.

Bilateral grid, an algorithm for rapidly computing the bilateral filter, which performs smoothing while preserving edges [Chen et al., 2007, Paris and Durand, 2009]. In essence, it creates a three-dimensional histogram of the image, where two dimensions correspond to x and y in image space, and the third corresponds to intensity. Applying a three-dimensional blur to this histogram has the effect of blurring across spatial dimensions where intensity is similar (i.e., where pixels are nearby in the third dimension) and preserving differences where there are sharp edges (where pixels are far apart in the third dimension).

With the exception of **edge** and **bilateral grid**, all of these algorithms can be mapped directly into line-buffered pipelines. **Bilateral grid** is somewhat unique in that it requires a mix of histograms, filtering, and trilinear interpolation for resampling. In the following chapters, we will describe how a domain-specific language can be built around the line-buffered pipeline machine model to express these algorithms, and show how appropriate restrictions in the model allow them to be easily mapped onto both CPU and FPGAs.

CHAPTER 4

Darkroom: A Stencil Language for Image Processing

This chapter describes *Darkroom*, a domain-specific language for image processing [Hegarty et al., 2014] built around line-buffered pipelines. By restricting the programming model, the compiler is able to rapidly schedule the code for hardware or software, producing efficient implementations for CPU, FPGA, and ASIC.

Figure 4.1 shows the three core features of the Darkroom programming model.

1. Line buffers, which store pixel values until enough have been loaded to service a stencil computation.

2. User-defined math computations, which produce one pixel per cycle. No control flow is allowed within the computations, but other arbitrarily complex expressions of standard mathematical operators are supported. Low-precision integer operations are preferred for hardware efficiency, but high-precision integer and floating-point operations are also allowed.

3. A directed acyclic graph (DAG) of line buffers and math computations. While many image processing applications (such as camera ISPs) are linear pipelines, others are expressed more effectively as a general DAG.

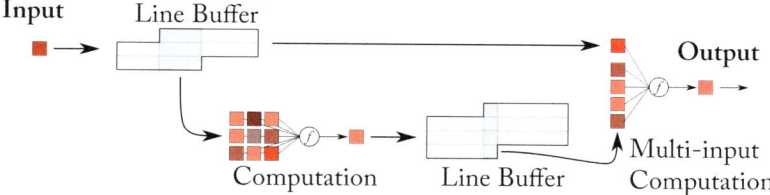

Figure 4.1: Line-buffered pipeline model that serves as the core abstraction in Darkroom.

Line buffers and math computations alternate in the DAG. Pointwise image processing operations do not need a line buffer and can be fused with the computations before or after them into a single math computation.

The line-buffered pipeline serves as the core intermediate representation within Darkroom. It is a particularly effective abstraction because it specifies all the key tradeoffs in image processing hardware without requiring the user to specify low-level details of the implementation, such as hardware clocks, physical interfaces, and so forth.

In the following sections, we describe the mechanics of the Darkroom language and how they facilitate programming with this model, and then show how the Darkroom compiler lowers this code down to hardware and software implementations.

4.1   LANGUAGE DESCRIPTION

Darkroom is a functional language, meaning that it consists only of "pure functions." In contrast to functions in procedural languages such as C or Java, the pure functions in Darkroom (and other functional languages) do not have side effects. That is, the output is computed from the inputs alone, with no changes to any state variables or memory. The benefit of this is that functions can be analyzed, optimized, and reordered far more easily than in procedural languages. The language only specifies *what* to compute, not how.

In Darkroom, a programmer specifies an image processing algorithm by defining intermediate images for each stage of the algorithm. Images at each stage of computation are specified as pure functions from 2D coordinates to the values at those coordinates, which we call *image functions*.

In Darkroom, image functions are declared using syntax loosely borrowed from lambda calculus, im(x,y). For example, a simple brightening operation applied to the input image I can be written as the function:

```
brighter = im(x,y) I(x,y) * 1.1 end
```

In contrast to languages like C, the loops over the x and y dimensions are implicit. The previous code example would have equivalent behavior to the following in C:

```
for(int y=0; y<HEIGHT; y++){
  for(int x=0; x<WIDTH; x++){
    brighter[x+Y*W] = I[x+y*W]*1.1f;
  }
}
```

Image functions are defined over all integer coordinates (x, y), though they can be explicitly *cropped* to a finite region using one of several boundary conditions. To implement stencil operations such as convolutions, Darkroom allows image functions to access neighboring pixels:

```
convolve = im(x,y) (1*I(x-1,y)+2*I(x,y)+1*I(x+1,y))/4 end
```

This has equivalent behavior to the following code in C:

```
for(int y=0; y<H; y++){
  for(int x=0; x<W; x++){
    convolve[x+Y*W] = (1*I[(x-1)+y*W] + 2*I[x+y*W] + 1*I[(x+1)+y*w])/4;
  }
}
```

To make it easier to express large stencil operations that would be cumbersome to write out explicitly, Darkroom also includes a map-reduce operator, with a syntax similar to a *for* loop:

```
boxFilter = im(x,y)
  map i=-5,5 j=-5,5 reduce(sum)
    I(x+i,y+j)
  end
end
boxFilterNorm = im(x,y) boxFilter/121 end
```

The map operator specifies an arbitrary number of indices with constant integer ranges (in this example, i=-5,5 j=-5,5). For each combination of values over the ranges, the expression inside the map-reduce is evaluated ("mapped"). This set of results is then merged using a "reduce" function, which can be chosen from a number of built-in reduce operations (sum, min, argmin, etc.). The map indices can be used anywhere within the body of the map-reduce expression, including the calculation of pixel coordinates. Because the index ranges are constant, the compiler can always analyze the bounds of the stencil.

The previous code using Darkroom's *map* would be equivalent to the following code in C:

```
for(int y=0; y<HEIGHT; y++){
  for(int x=0; x<WIDTH; x++){
    int sum = 0;
    for(int j=-5; j<=5; j++){
      for(int i=-5; i<=5; i++){
        sum += I[(x+i)*(y+j)*W];
      }
    }
    boxFilterNorm[x+Y*W] = sum/121;
  }
}
```

To support operations like image warps, where the pixel offsets are functions rather than constants, Darkroom has an explicit gather operator. Gathers allow dynamically computed indices, as long as the indices are explicitly bounded within a compile-time constant distance from the current (x, y) position. For example, we could create two functions warpVX and warpVY, which return integer offsets based on some patch of image data. A common pattern for this is to use argmin or argmax to find the pixel offset that maximizes some metric. Then we can apply the gather operation:

```
1 warp = im(x,y) gather(I, 4, 4, warpVX(x,y), warpVY(x,y)) end
```

The values of `warpVX` and `warpVY` are clamped to the range $[-4, 4]$, ensuring that the range of the gather is bounded and statically analyzable.

Darkroom's gather would behave equivalently to the following code in C:

```
1 for(int y=0; y<HEIGHT; y++){
2   for(int x=0; x<WIDTH; x++){
3     warp[x+Y*W] = I[(x+warpVX[x+y*W])+(y+warpVY[x+y*W])*W];
4   }
5 }
```

As these examples illustrate, Darkroom makes the following restrictions to fit within the line-buffered pipeline model:

1. Image functions can only be accessed (1) at an index $(x + A, y + B)$ where A, B are constants, or (2) with the explicit gather operator. Affine indices like `I(x*2,y*2)` are not allowed. This means that every stage produces and consumes pixels at the same rate, a restriction of line-buffered pipelines.

2. Image functions cannot be recursive (e.g., computing an integral image), because this could force a serialization in how the image is computed. This makes it impossible to implement inherently serial techniques inside a pipeline.

4.2   A SIMPLE PIPELINE IN DARKROOM

Now that we have introduced the basics of the Darkroom language, let us apply these ideas to a simple example, the **unsharp** algorithm. Implementing the 2D blur as separate 1D passes, we could write the pipeline as:

```
1 bx = im(x,y) (I(x-1,y) + I(x,y) + I(x+1,y))/3 end
2 by = im(x,y) (bx(x,y-1) + bx(x,y) + bx(x,y+1))/3 end
3 difference = im(x,y) I(x,y) - by(x,y) end
4 scaled = im(x,y) 0.5 * difference(x,y) end
5 sharpened = im(x,y) I(x,y) + scaled(x,y) end
```

The final three image functions—`difference`, `scaled`, and `sharpened`—are pointwise operations, so the whole pipeline can be collapsed into two stencil stages, as shown in the code below and illustrated in Figure 4.2:

```
1 S1 = im(x,y) (I(x-1,y) + I(x,y) + I(x+1,y))/3 end
2 S2 = im(x,y)
3         I(x,y) + 0.5*(I(x,y)-(S1(x,y-1) + S1(x,y) + S1(x,y+1))/3)
4       end
```

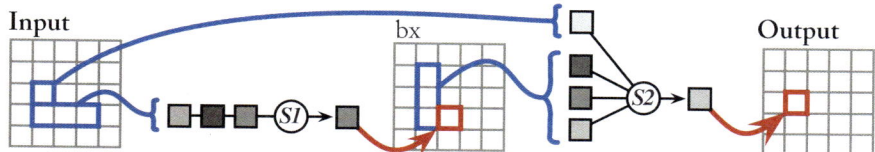

Figure 4.2: Intermediate buffers and stencil sizes of an unsharp mask pipeline after pointwise operations have been fused into single math computations.

The pipeline cannot be collapsed any further without changing the stencils of the individual computations. Notice that this is not a linear pipeline, but a general DAG of operations communicating through stencils. The final sharpened result is composed of stencils from both the original input image and the horizontally blurred intermediate.

4.3   OPTIMAL SYNTHESIS OF LINE-BUFFERED PIPELINES

By design, the Darkroom language allows the user to declare what operations must occur in their image-processing pipeline, but does not specify how to compute the pipeline in hardware, whether CPU or FPGA. In this section, we describe the algorithm that automatically transforms a Darkroom program into a line-buffered pipeline, which can be synthesized into hardware. The key problem in compiling Darkroom code to a line-buffered pipeline is deciding when to perform each operation such that the size of the line buffers is minimized, a process we will refer to as *scheduling*. In image-processing ASIC designs, the line buffers typically consume more die area than the compute [Brunhaver, 2015], so it is crucial that our compiler not increase the size of on-chip buffering more than is necessary. In the remainder of this section, we will show that it is possible to formulate line buffer minimization as an Integer Linear Programming (ILP) problem, which will allocate buffers optimally in less than a second of runtime.

4.3.1   GENERATING LINE-BUFFERED PIPELINES

One way to think about the scheduling problem is to consider a translation from image space into a time sequence. If the image is streamed in, pixel by pixel, then we can map operations in image space into operations on a temporal schedule. This is best explained by an example.

Figure 4.3a shows Darkroom code for a simple 1D convolution of an input In with a constant kernel k_i. To translate this code into a schedule, we compute pixel i at time $t = i$. Following this rule, we will schedule the pipeline such that at time $t = 0$, we calculate In(0) and Out(0), and at time $t = 2$ we calculate In(2) and Out(2), illustrated in Figure 4.3b. Reads of values computed in the same cycle such as In(0) can be wired directly to their consumers. Values from the past such as In(x - 2) are stored as entries in an N pixel shift register, a line buffer. Figure 4.3c shows the line-buffered pipeline that results from our simple example. While this scheduling algorithm works correctly for this simple example, we will see that it does not work well for all pipelines.

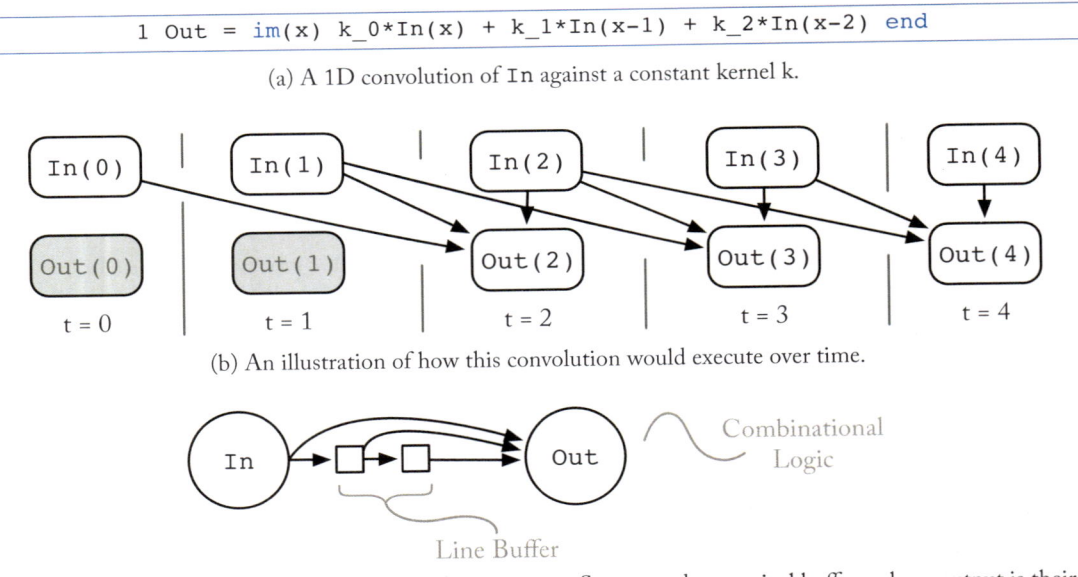

```
1 Out = im(x) k_0*In(x) + k_1*In(x-1) + k_2*In(x-2) end
```

(a) A 1D convolution of In against a constant kernel k.

(b) An illustration of how this convolution would execute over time.

(c) A line-buffered pipeline that implements this execution. Square nodes are pixel buffers whose output is their input from the *previous* time step. Circular nodes represent computations. Two pixel buffers are required because Out accesses values of In two cycles in the past. This collection of pixel buffers forms a *line buffer*.

Figure 4.3: Direct translation of a simple convolution stencil into a line-buffered pipeline.

This 1D scheduling algorithm can also handle two-dimensional stencils by reducing them to a one-dimensional access pattern. We will assume that the image is stored in memory in row-major, scan-line order, which is the most common read-out pattern for image sensors. Given a fixed line size L, accesses $f(x + c_1, y + c_2)$ are replaced with $f'(x' + c_1 + L * c_2)$ where $x' = x + L * y$ is the current pixel in the stream. This rewrite results in the same accesses to memory but fits within the 1D restriction. For the remainder of the section, we will assume that the input code has already been transformed in this way, without loss of generality.

So far, we have only handled stencils that access data from the current time step or the past. In signal processing these are referred to as *causal* filters. Figure 4.4a shows an example program where this is not the case. It performs a Richardson-Lucy deconvolution, taking as input the latent estimate of the deconvolved image Lat, the blurred input image Obs, and constants k_i that describe the point spread function of the blur. It calculates the relative error Rel of the latent estimate and produces an improved latent estimate LatN.

If we directly translate this code into a pipeline as we did before, at time t, we would calculate the value of each intermediate f at position t. For instance, when $t = 0$, we would calculate `Lat(0)` and `Est(0)`. But for this example, we cannot compute `Rel(0)` since it depends on `Lat(1)`, which is not calculated until $t = 1$. This problem is a read-after-write hazard, illustrated in Figure 4.4b.

We can transform non-causal pipelines like this into causal ones by shifting the time at which values are calculated relative to others. We first introduce the shift operator and then discuss how to choose shifts that ensure causality and minimize line buffering.

4.3.2   SHIFT OPERATOR

In our example, if we want to ensure `Rel` only relies on *current* or *previous* values of its input, we can *shift* it in time to eliminate the hazard. We define a shift operator for an integer shift s:

$$f_s(x) = f(x - s)$$

That is, at time step $t = s$, f_s will produce the value $f(0)$. We can now replace uses of a value with the equivalent shifted value. For instance, we replace `Rel` with `Rel_1` and adjust the offsets:

```
Rel_1 = im(x) Obs(x-1) / (k_0*Lat(x-2)+k_1*Lat(x-1)+k_2*Lat(x)) end
```

Now all uses of `Lat` are previous values. We also need to adjust all the uses of `Rel` to be in terms of Rel_1:

```
LatN = im(x) Lat(x)*(k_2*Rel_1(x)+k_1*Rel_1(x+1)+k_0*Rel_1(x+2)) end
```

The effect of this shift is visualized in Figure 4.4c. Note that in this case it introduced an additional hazard. We can also shift `LatN` by 2, which results in this modified program that contains no hazards:

```
Rel_1   = im(x) Obs(x-1) / (k_0*Lat(x-2)+k_1*Lat(x-1)+k_2*Lat(x)) end
LatN_2 = im(x) Lat(x-2)*(k_2*Rel_1(x-2)+k_1*Rel_1(x-1)+k_0*Rel_1(x)) end
```

Figure 4.4d illustrates how this pipeline will execute, and Figure 4.4e shows the result of translating it into a line-buffered pipeline. As before, values accessed at the same time are piped directly to each other, while values accessed in the past are implemented by inserting buffers. Calculating the original function (`LatN`) given a shifted pipeline that executes it (`LatN_2`) simply requires changing the indices that are calculated each cycle, for example, `LatN_2(2)` at $t = 0$ instead of `LatN(0)`. Similarly, evaluating shifted leaf nodes such as inputs from DRAM or the sensor simply requires shifting which address is read.

4.3.3   FINDING OPTIMAL SHIFTS

Despite being correct, the pipeline in Figure 4.4e is not optimal: there is an unnecessary line buffer after `Obs` that would disappear if we choose to shift `Obs` by 1. To create an optimal pipeline, we must choose shifts that *both* ensure causality *and* minimize line buffer size.

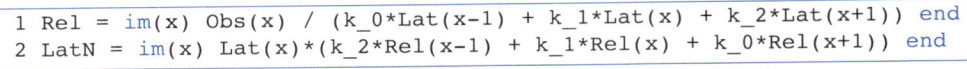

```
1 Rel = im(x) Obs(x) / (k_0*Lat(x-1) + k_1*Lat(x) + k_2*Lat(x+1)) end
2 LatN = im(x) Lat(x)*(k_2*Rel(x-1) + k_1*Rel(x) + k_0*Rel(x+1)) end
```

(a) Code for a 1D Richardson-Lucy deconvolution.

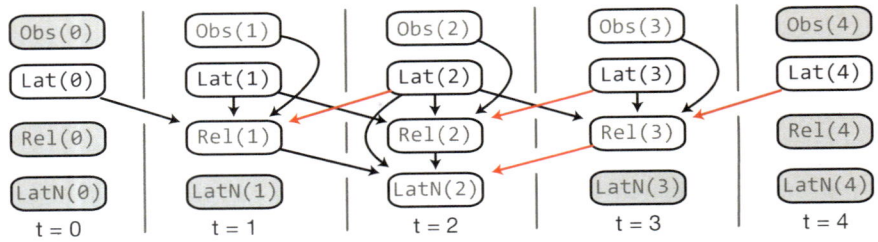

(b) Scheduling this code naively results in a non-causal pipeline—it accesses into the future (shown in red).

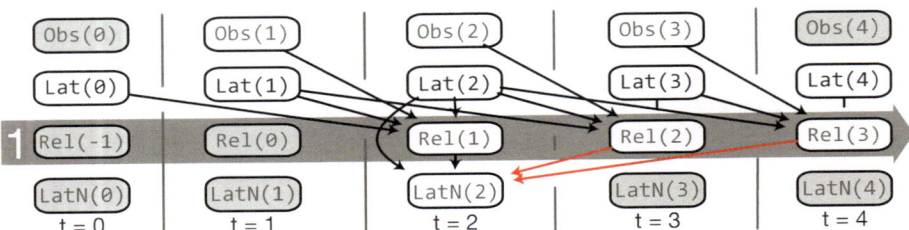

(c) We can shift a value temporally to eliminate hazards. Here we have shifted Rel by 1.

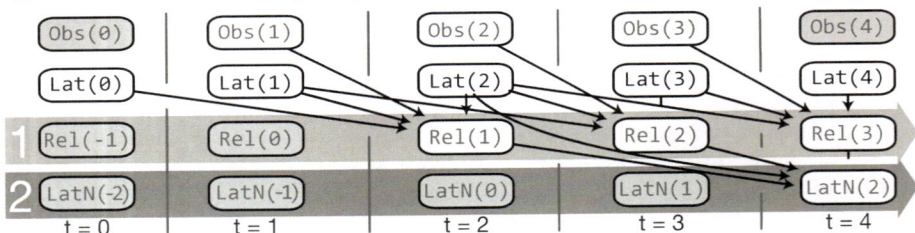

(d) This operation can introduce hazards later in the pipeline, which can be fixed by later shifts.

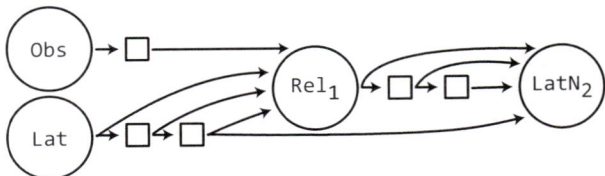

(e) After eliminating hazards, we can construct a pipeline using line buffers to store previous values. Here is a correct pipeline for deconvolution.

Figure 4.4: Applying shifts makes a non-causal pipeline realizable.

The general case is complicated by the fact the program may have multiple inputs and multiple outputs (e.g., an RGB image and a separately calculated depth map). Furthermore, individual line buffers are not always the same size. For instance, some values may be one-byte grayscale while others might be three-byte RGB triples. Figure 4.5 shows an example of where different sized outputs can produce different scheduling decisions.

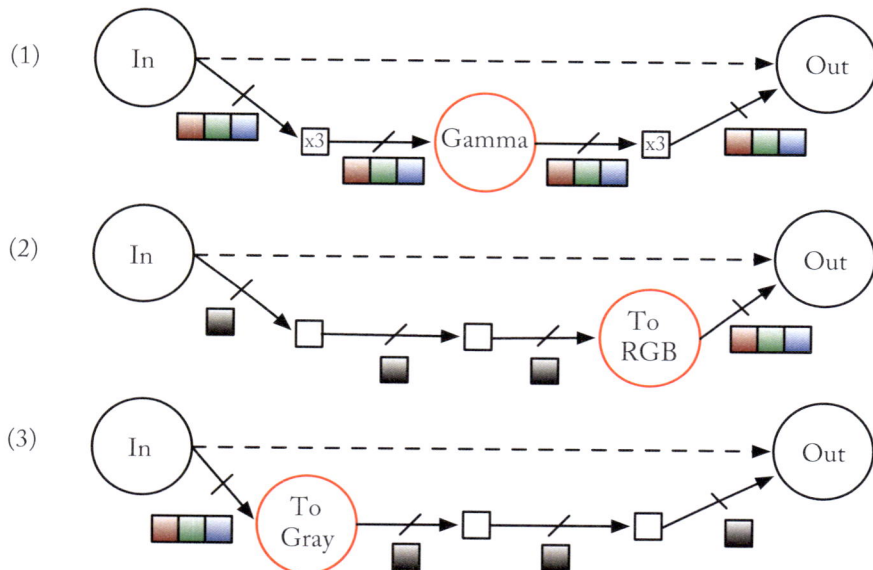

Figure 4.5: Given some path between In and Out of a fixed length, the optimal placement of line buffers through the red node to match the fixed path depends on the size of the pixels entering and leaving it. In (1) it produces the same amount of data and the choice does not matter, but in (2) it produces more data so the buffers should be placed before it, while in (3) it reduces data so the buffers should be placed after it.

We can formulate this optimization as an integer linear programming problem. Let F be the set of image functions, p and c be image functions in F, and d be a stencil offset. If the value $p(x + d)$ is evaluated in the process of evaluating $c(x)$, we generate a *use* triple (c, p, d). Let U be the set of all uses in a program. For instance, the program:

```
Out = im(x) In(x - 1) + In(x) + In(x + 1) end
```

will result in the following values for F and U:

$$F = \{Out, In\}$$

$$U = \{(Out, In, -1), (Out, In, 0), (Out, In, 1)\}$$

Also, since the size of a pixel data type varies, we extract b_f, the pixel size of each image in bytes during type checking.

For each image function f, we want to solve for its shift s_f to ensure causality and minimize line buffer size. For each use (c, p, d) we calculate the number of delay buffers $n_{(c,p,d)}$ needed to store the value between when it is produced and when it is consumed as:

$$n_{(c,p,d)} = s_c - s_p - d.$$

A negative number of delay buffers indicates a non-causal pipeline, so, to address causality, for each use we add the following constraint to the integer linear program:

$$n_{(c,p,d)} \geq 0.$$

The line buffer for an image function f can be shared by all of its consumers, so for each image function $f \in F$, we calculate the size of its line buffer as the maximum number of delays needed by any consumer scaled by the pixel size, $(\max_{(c,f,d) \in U} n_{(c,f,d)}) * b_f$. The total size of the line buffers in the entire pipeline S is the sum of the line buffers of all producers:

$$S = \sum_{p \in F} (\max_{(c,p,d) \in U} n_{(c,p,d)}) * b_p.$$

We use S as the objective to minimize in the integer linear program. Our scheduling formulation ends up being equivalent to the problem of minimizing register counts in circuit retiming literature. Minimizing register counts can be formulated as min-cost flow, which has a polynomial time solution [Leiserson and Saxe, 1991]. In practice, however, we have found that implementing our scheduling formulation directly using ILP is convenient and gives sufficient performance.

4.4   IMPLEMENTATION

After generating an optimized line-buffered pipeline, our compiler instantiates concrete versions of the pipeline as ASIC or FPGA hardware designs, or code for CPUs (Figure 4.6). The Darkroom compiler is implemented as a library in the Terra language [DeVito et al., 2013] that provides the im operator. When compiled, Darkroom programs are first converted into an intermediate representation (IR) that forms a DAG of high-level stencil operations. We perform standard compiler optimizations such as common sub-expression elimination and constant propagation on this IR. A program analysis is done on this IR to generate the ILP formulation of line buffer optimization, described in the previous section. We solve for the optimal shifts using an off-the-shelf ILP solver [lp_solve contributors, 2010], and use them to construct the optimized pipeline. It converges to a global optimum in less than a second on all of our test applications. The optimized pipeline is then fed as input to either the hardware generator, which creates ASIC designs and FPGA code, or the software compiler, which creates CPU code.

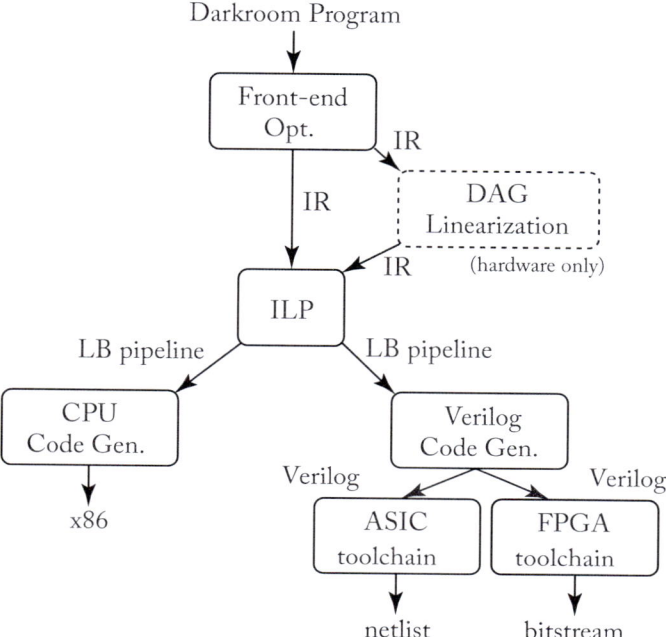

Figure 4.6: The stages of the Darkroom compiler.

4.4.1 ASIC AND FPGA SYNTHESIS

Our hardware generator implements line buffers as circularly addressed static RAMs (SRAMs) for ASIC or block RAMs (BRAMs) for FPGA. Each clock cycle, a column of pixel data from the line buffer shifts into a 2D array of registers. These registers save the bandwidth of reading the whole stencil from the line buffer every cycle. The user's image function is implemented as combinational logic from this array of shift registers, writing into an output register (Figure 4.7). We add small FIFO buffers to absorb stalls at the input and output of each line buffer.

Instantiating line buffers as real SRAMs presents additional difficulties beyond those present in our abstract pipeline model. First, SRAMs and BRAMs are only available in discrete sizes, each with different costs. Second, they have limited bandwidth, preventing multiple image functions reading from them simultaneously. To simplify these issues, the current version of Darkroom's hardware generator only supports programs that are straight pipelines with one input, one output, and a single consumer of each intermediate.

Many Darkroom programs we have written contain a DAG of dependencies, where image functions have multiple inputs and multiple outputs. In order to support these programs in our hardware implementation, we first translate the Darkroom program into an equivalent Darkroom program that is a straight pipeline. This process is described in Figure 4.8. While

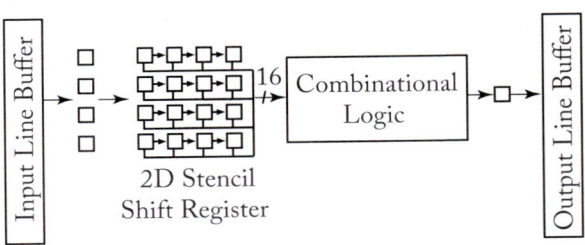

Figure 4.7: Darkroom's hardware generators synthesize stages in a line-buffered pipeline using a common microarchitectural template. Columns of pixels are shifted from the input line buffer into a stencil register, processed through arbitrary computation in the data path, and pushed onto the output line buffer. According to the constraints of our programming model, the data path only has access to constants and the current contents of the stencil register.

this process always produces semantically correct results, the merging of nodes in the programs can create larger line buffers than what could be achieved with a hardware implementation that supported DAG pipelines. In the future, multi-ported buffers would eliminate this restriction.

Following DAG linearization, we use Genesis2, a Verilog meta-programming language [Shacham et al., 2012] to elaborate the topology into a SystemVerilog hardware description for synthesis. We verify functionality of the hardware description using Synopsys VCS G-2012.09, which also produces the activity factor required for ASIC power analysis. The ASIC design is synthesized and analyzed using Synopsys Design Compiler Topographical G-2012.06-SP5-1 for a 45 nm cell library. The FPGA design uses Synopsys Synplify G-2012.09-SP1 to synthesize the design and Xilinx Vivado 2013.3 to place and route the design for the Zynq XC7Z045 system-on-a-chip on the Zynq ZC706 demo board. The FPGA performance is measured on the demo board using a custom Linux kernel module.

4.4.2 CPU COMPILATION

We can compile efficient CPU code from the same intermediate representation. Our CPU compiler implements the line-buffered pipeline as a multithreaded function. To enable parallelism, we slice the output image into multiple strips and compute each strip on a different core. Intermediates along strip boundaries must be recomputed, but this is a small price to pay to gain use of all the available cores.

Within a thread, the code we generate is a software realization of the line-buffered pipeline model. Line buffers are implemented using a small block of memory simulating a scratchpad, which is held in cache by restricting memory accesses to this block and issuing non-temporal writes for the output results, following the technique of Gummaraju and Rosenblum [2005].

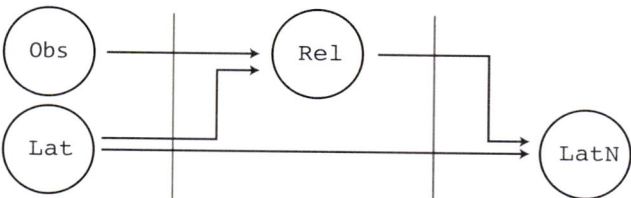

1. Group IR nodes by distance from the input:

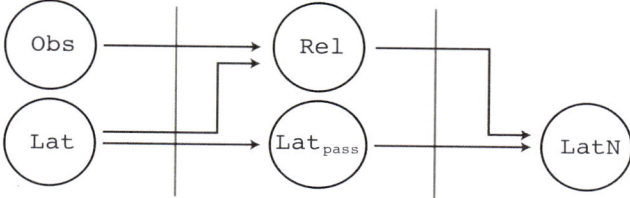

2. Add passthrough nodes whenever an edge crosses a stage:

3. Merge values in each stage, producing larger pixel widths:

The transform results in the following code:

```
RelLatPass = im(x)
    {ObsLat(x)[0] / ( k0 * ObsLat(x-1)[1] +
                      k1 * ObsLat(x)[1] +
                      k2 * ObsLat(x+1)[1]),
      ObsLat(x)[0]}
end
LatN = im(x)
    RelLatPass(x)[1] * ( k2 * RelLatPass(x-1)[0] +
                         k1 * RelLatPass(x)[0] +
                         k0 * RelLatPass(x+1)[0])
end
```

The code is then transformed into a pipeline:

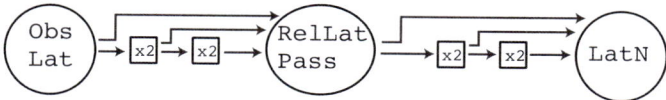

Buffers contain merged values, so they are twice as large.

Figure 4.8: Converting a generic program DAG into a linearized pipeline by merging nodes.

One simple compilation schedule would be to have the thread's main loop correspond to one clock cycle of the hardware, with the individual operations scheduled in topological order to satisfy dependencies. A simple two-stage pipeline with stages *S1* and *S2* would be scheduled:

> **for** each pixel x, y
> > compute $S1(x, y)$ // *Store pixel in line buffer*
> > compute $S2(x, y)$ // *Load stencil of S1 from line buffer*

However, one downside of this schedule is that the entire set of line buffers will often exceed the size of the nearest (and fastest) level of the CPU cache. To mitigate this, we block the computation at the granularity of image lines. In this schedule, the main loop calculates one line of each stencil operation, and the line buffers are expanded to the granularity of whole lines:

> **for** each line y
> > **for** each pixel x in line of *S1*
> > > compute $S1(x, y)$ // *Store pixel in line buffer*
> >
> > **for** each pixel x in line of *S2*
> > > compute $S2(x, y)$ // *Load stencil of S1 from line buffer*
> >
> > rotate line buffers

With image-line blocking, the entire set of input lines that an image function consumes is often still resident in the closest cache. In addition, this blocking scheme reduces register spills in the inner loop by reducing the number of live induction variables.

To exploit vector instructions available on modern hardware, we vectorize the computation within each line of each stage. For intermediates, we store pixels in struct-of-array form to avoid expensive gather instructions.

The compiler is implemented using Terra to generate low-level CPU code including vectors and threads, which is compiled and optimized using LLVM [Lattner and Adve, 2004].

4.5   EVALUATION

To evaluate Darkroom, we implemented most of the image-processing applications described in Chapter 3.

Mapping **campipe** to Darkroom is straightforward: it is a linear pipeline of stencil operations, each of which becomes an image function. **Campipe** is non-trivial, however, due to its size: it is 472 lines of Darkroom code, which must be scheduled and compiled into hardware or software.

Corner is a simple DAG of operations that also maps nicely to Darkroom. The resulting corner points are superimposed on the input image for display. **Edge** is similar, except that the final edge propagation step is not bounded by a stencil. We work around this by implementing a series of ten single-pixel propagation steps. This means that a lower-threshold edge will only propagate ten pixels, but in practice, this limit makes very little difference in the behavior of the algorithm.

Deblur is a computationally intense iterative algorithm, which we use as a stress test of our system. We unrolled **deblur** to eight iterations, which was the maximum size our hardware synthesis tools could support.

Stereo makes use of Darkroom's bounded gather to rectify the images, with the bound determined by the largest offset in the rectification map. It is by far the most computationally intensive pipeline we tested.

Flow normally is run on a multi resolution pyramid. This is not supported by Darkroom, because this implies that the entire image is read in (and buffered) before any output can be produced. However, a single level of the pyramid can be implemented as a line-buffered pipeline, and so our test includes a single-scale version that operates only at the finest resolution.

All of the applications use fixed-point arithmetic for efficiency. For the purposes of evaluation, all pipelines were tested independently with whatever input images were appropriate. On a real camera, the image-processing algorithms would likely be fed the output from **campipe**.

4.5.1 SCHEDULING FOR HARDWARE SYNTHESIS

With Darkroom, we scheduled, compiled, and synthesized **campipe**, **edge**, **corner**, and **deblur** into designs for both FPGA and ASIC. Table 4.1 shows the throughput and resource utilization on a Xilinx Zynq 7045, a mid-range FPGA/ARM SoC platform costing $1000. The designs all run faster than 120 megapixels/second, which is sufficient to process 1080p video at 60 frames per second.

Table 4.1: Darkroom compiles our applications well within the resource limits of a mid-range FPGA while delivering enough performance from all pipelines to process 1080p/60 video in real-time. Resource utilization is reported as a percentage of the available resources on a Xilinx Zynq 7045. In practice, this platform provides enough resources to compile much larger pipelines, implementing multiple vision and image-processing algorithms simultaneously in real time.

Pipeline	MPix/s	Resource Utilization		
		LUTs	BRAMs	DSPs
campipe	143	26%	7.5%	6.3%
edge	131	7.2%	6.5%	3.9%
corner	146	5.6%	4.8%	3.8%
deblur	125	49%	59%	50%

An FPGA has fixed computing resources, unlike an ASIC (where more compute simply requires more silicon area) and a CPU (where more compute requires more time). The key resources are look up tables (LUTs) used to implement combinational logic, block RAMs (BRAMs) used for line buffers, and the "DSP" blocks that implement 24-bit integer arithmetic

(DSPs). The applications in Table 4.1 use only a fraction of the total resources, leaving room for additional computation and storage. However, an application as large as **stereo** does not fit on the FPGA, and we'll explore how we can handle this in the next chapter.

Table 4.2 summarizes the energy and area breakdown for the ASIC designs, simulated on a leading foundry's 45 nm process. The designs run at rates between 940 and 1040 megapixels/second (approximately equivalent to processing 4 K video at 120 frames/second), with power budgets around 200 mW. Both the silicon area and power results are similar to hand-designed commercial ISP pipelines [Aptina, 2016].

Table 4.2: Darkroom compiles our applications into efficient ASICs, using a chip area of 0.3–2.6 mm^2 and energy efficiency of 165–1280 pJ/pixel for compute, and < 3.2 nJ/pixel including the communication with DRAM. Processing 720p, 30 frame per second footage requires < 90 mW for all pipelines (including DRAM energy). This is similar to commercially available camera modules with ISPs, which use 550 mW total power for 720p, 30 frame per second capture [Aptina, 2016].

Pipeline	Energy (pJ/pix)		Area (mm^2)	
	Compute	DRAM	Total	LB
campipe	224	1,360	0.36	0.19
edge	202	1,040	0.29	0.17
corner	165	1,000	0.25	0.12
deblur	1,280	1,920	2.56	1.10

The important thing to observe is that despite the amount of computation these pipelines perform, the energy required to fetch the pixels from DRAM is still larger—by as much as a factor of five. Computation really is all about memory, even for these very dense high-locality applications.

Given that, there is room to build much deeper, more complex Darkroom pipelines without appreciably increasing the overall energy budget. In several of these applications, a 10× increase in energy spent on computation would only double the total energy budget.

Darkroom is built around the philosophy that efficiently using on-chip buffering is essential to minimizing expensive off-chip DRAM traffic. In practice, two effects limit the actual buffering efficiency of our synthesized hardware designs relative to the optimal line-buffering schedule computed by integer linear programming:

1. Our hardware generators currently require the acyclic graphs to be flattened into linear pipelines to simplify synthesis, introducing additional buffering where earlier values are passed through intermediate stages.

2. Our FPGA hardware generator adds additional overhead due to the coarse granularity of BRAM allocation, combined with several simplifying assumptions in our logic generator.

Our ASIC backend synthesizes buffers with very little overhead, so inefficiency in our ASIC designs comes almost exclusively from linearization (1). FPGA designs can add significant additional overhead (2). In practice, for the pipelines we studied in hardware, linearization increases buffering by at most 2.9× above optimal, while FPGA overhead increases buffering up to 3.2×. There are opportunities to improve BRAM allocation in our FPGA generator, but it has not been the limiting factor for these applications.

4.5.2   SCHEDULING FOR GENERAL-PURPOSE PROCESSORS

Using exactly the same source code, Darkroom is able to produce well-optimized CPU code that greatly outperforms a naive C++ implementation and which matches optimized code from Halide (which we'll describe in more detail in the next chapter). A summary of the results on an x86 CPU (a 4-core 3.5 GHz Intel Core i7 3770) is shown in Table 4.3. **Campipe** was benchmarked on a 7-megapixel raw image, **flow** on a 1080p video, **stereo** on a 480p stereo video, and **edge**, **corner**, and **deblur** were each benchmarked on 16-megapixel still images.

Table 4.3: Darkroom programs compiled to a quad-core x86 CPU deliver throughput sufficient to process 720p, 24 frames per second video on most applications, or as high as 1080p, 60 frames per second for **corner**. **Stereo** delivers a much lower pixel rate than other applications because it performs dramatically more arithmetic; its performance is proportional to the difference in arithmetic per pixel. Allocated line buffer storage is near optimal in all cases.

Pipeline	Throughput (Mpix/sec)		Buffering (kB)	
	1 Core	4 Cores	Optimal	Achieved
campipe	7.5	24	427	436
edge	12	34	224	228
corner	78	148	108	110
deblur	4	14	1,596	1,622
flow	7.8	22	1,404	1,431
stereo	0.067	0.25	108	122

As we would expect, throughput varies based on the computational intensity of the pipeline: roughly 30 megapixels/sec on 4 cores for each of **campipe**, **edge**, **deblur**, and **flow**, as high as 148 megapixels/sec for the comparatively simple **corner** pipeline, and as low as 0.25 megapixels/sec for **stereo**.

The CPU consumes approximately 85 W of power under load, which corresponds to approximately 500–5000 nJ/pixel (for applications other than **stereo**). This is 2–3 orders of magnitude more than equivalent ASIC hardware plus DRAM, which lines up with the values we laid out in Chapter 1.

Unlike the hardware generator, which incurs significant overhead from linearizing the pipeline, the CPU version is fairly efficient with buffering. It does introduce modest overhead for the sake of speed (e.g., enlarging buffers for better vector alignment), but achieved buffering is always within 13% of optimal, and generally less than 2%.

For **campipe**, we compared Darkroom to a reference implementation written as clean C. Our reference code has no multithreading, vectorization, or line buffering. Enabling these optimizations by reimplementing it in Darkroom yielded a 7× speedup, with source code of similar complexity. Of this speedup, 3.5× comes from multithreading, and 2× comes from vectorization.

Darkroom performs similarly to Halide, another image processing DSL with a more flexible scheduling model. On the **deblur** application (for which there was a convenient reference implementation), Darkroom performs 1.1× faster than Halide. Here, the advantage of Darkroom is not speed but rather the ability to produce a high-performance implementation automatically, without direction from the developer. As we'll see in the next chapter, Halide opens up a much wider range of scheduling possibilities, but this requires that the developer specify at a high level how to schedule the algorithm.

4.6   SUMMARY

In this chapter, we have described how high-level Darkroom code can be synthesized into a line-buffered pipeline, which is mapped to either CPU or a hardware design for FPGA or ASIC. By providing algorithm-level constructs, rather than implementation-level control, the Darkroom language can target these completely different platforms with competitive performance.

A key restriction is that the pipelines described so far all operate in lockstep at 1 pixel per cycle. A great many interesting image processing applications fit into this programming model, such as camera ISP pipelines and deconvolution. However, many do not, such as those that upsample or downsample the image, or extract a sparse set of features. It is also often advantageous to make space-time tradeoffs by duplicating computation logic to increase throughput or reusing it over multiple clock cycles to save silicon area. In the next chapter, we explore the design and implementation for a similar system where this restriction is relaxed.

CHAPTER 5

Programming CPU/FPGA Systems from Halide

In the previous chapter, we showed how the Darkroom DSL is designed from scratch to map high-level image processing algorithms to hardware. By restricting the language to stencil computations connected with line buffers, the Darkroom compiler is able to lower designs to hardware and software, with no implementation detail provided by the programmer.

This chapter describes a second approach to creating hardware and software designs from a DSL. Rather than designing a new language from scratch, we begin with Halide, a popular open-source image-processing DSL. Halide is a more general language than Darkroom and can express algorithms which do not fit into the strict line-buffered pipelines we have described so far. Using Halide has the additional benefit that developers can port existing Halide applications to our new CPU/FPGA target without rewriting their algorithms.

Halide makes this high portability possible by separating the essential computation to be performed (the *algorithm*) from the order in which it is done (the *schedule*). To port a Halide application to a different machine target, the programmer only needs to change the schedule for workload mapping and performance tuning, and the Halide compiler guarantees the correctness of the implementation of the algorithm.

While Halide is able to express a wider range of image-processing algorithms, this increased generality means that the scheduling problem is much harder. We address this by extending Halide's scheduling language and designing compiler techniques to allow users to schedule portions of image-processing pipelines as line-buffered pipelines and map them to FPGA. By comparison, Darkroom has only *algorithm*; a *schedule* is unnecessary because Darkroom can always compute an optimal schedule for any algorithm that can be expressed.

Furthermore, the Halide framework is able to quickly program the whole CPU/FPGA SoC in one pass, creating not only FPGA accelerator kernels, but also the host CPU code and proper driver configurations to access the FPGA hardware.

We begin the chapter with a brief explanation of the key features of the Halide language and will show in Section 5.2 how the language features facilitate the mapping of algorithms onto heterogeneous systems. Next, Section 5.3 describes our compiler system that implements these extensions to produce blended CPU/FPGA designs. The final section describes an FPGA test platform where we evaluate the efficacy of the system.

5.1   THE HALIDE LANGUAGE

As discussed earlier, the design space for implementing an image-processing algorithm is huge, due to the choices for vectorization, tiling, multithreading, and so on. Given that most of these options require restructuring the code, finding the most efficient implementation for an application is difficult and time consuming.

The Halide language was invented to tackle this problem. The key idea is to decouple the computation to be performed (the *algorithm*) from the order in which it is done (the *schedule*). The developer specifies the function to compute, and the compiler can easily synthesize loops that calculate the desired result. Separately, the developer also specifies the desired loop nesting, which enables the compiler to split, fuse, and reorder the computation loops, and apply various optimizations to the code.

Because schedules can be specified with only a few lines of code, it is easy to experiment with new schedules. And because the algorithm and schedule are separated, performance tweaks do not change the algorithm code, and thus optimizing the schedule is guaranteed not to break the correctness of the algorithm.

To achieve this separation, Halide represents algorithms in a pure functional form, much like Darkroom, although with a slightly different syntax. For example, a separable 3×3 box blurring filter can be expressed as a chain of two functions in x, y as follows:

```
1  Var x, y;
2  Func blury, blurx;
3  Image input;
4  blury(x, y) = (input(x, y-1) + input(x, y) + input(x, y+1)) / 3;
5  blurx(x, y) = (blury(x-1, y) + blury(x, y) + blury(x+1, y)) / 3;
```

In this form, functions are data parallel by construction, and the dataflow of functions can be extracted statically.

The order and the range of functions to be evaluated are treated as optimization choices and are specified using the language of schedules. This is based on a set of loop transformation concepts, including loop splitting (computing a large loop as a small fine-grained loop nested inside a coarse one), fusion (the reverse of splitting), reordering (swapping inner and outer loops in a nest), and tiling (splitting over multiple dimensions simultaneously).

The language provides *scheduling primitives* for applying these transformations on each function and defining the granularity with which to interleave the computation of each function. For example, a CPU-optimized schedule for the blurring filter can be written in Halide as follows.

```
1  blury.tile(x, y, xi, yi, 256, 32)
2        .vectorize(xi, 8).parallel(y);
3  blurx.compute_at(blur_y, x).vectorize(x, 8)
```

The `tile` primitive partitions the computation of the output image into 256×32 chunks, and the `compute_at` primitive specifies that the intermediate image `blurx` is computed for each chunk to improve cache locality. To exploit the data-parallelism, the `vectorize` and `parallel` directives are used to mark loops to be executed using SIMD datapaths and multi core in the CPU.

In addition to expressing loop transformations for image-processing algorithms on different machine architectures, Halide's schedules can also specify the a portion of an algorithm to be executed on accelerator devices (e.g., GPUs or programmable DSPs). The Halide compiler can then analyze the dataflow between the host and the devices, and automatically exploit the concurrency of computation and data transfer and efficient memory management (e.g., late allocation and early free, zero-copy buffers, etc.). For example, we can offload one of the blurring stages, `blury`, on a GPU device using the following schedule.

```
1  blury.tile(x, y, xo, yo, xi, yi, 8, 8)
2      .gpu_blocks(xo, yo)
3      .gpu_threads(xi, yi);
4  blurx.compute_root();
```

The schedule tiles `blury` into 8×8 chunks. Each chuck uses a parallel GPU block, while each pixel in a chunk uses a GPU thread (in CUDA terminology).

In the above example, the `blurx` image is still computed on the host CPU via the `compute_root` schedule. However, the device memory management and the data transfers between the host and device are automatically handled during the Halide code generation without any additional user code. This is possible because the computations of both the host and the device are defined in a unified language (a Halide algorithm) and the algorithm mapping (a Halide schedule) is statically given.

5.2   MAPPING HALIDE TO HARDWARE

Although Halide algorithms present functional specifications that are architecture independent, its scheduling primitives are tailored for data-parallel architectures (e.g., CPUs and GPUs), and are insufficient for describing the hardware architecture for the FPGA target and the orchestration of the generated FPGA accelerator on a CPU/FPGA platform. Our task is to extend Halide's scheduling to cover the new target. Specifically, the schedule should include:

- the scope and interface of the hardware accelerator pipeline;

- the granularity of the accelerator launch task, in other words, the portion of the output image block the hardware produces per launch;

- the amount of parallelism implemented in the hardware datapath, which affects the throughput of each pipeline stage;

- the allocation of buffers, specifically line buffers, that optimally trades storage resources for less re computation; and

- the number of delay register slices needed to match varying computation latencies.

Many hardware scheduling choices have analogues in CPU scheduling, and existing Halide primitives can be used to describe them. For example, both CPU and hardware schedules must describe computation order and memory allocation. In such cases, we reuse as many of the existing primitives as possible. Ultimately, we are able to achieve efficient hardware mapping and hybrid CPU/accelerator execution using only two new primitives and a bit of syntactic sugar.

The additional scheduling primitives are best explained in the context of an example. Figure 5.1a shows a simple **unsharp** mask filter implemented in Halide. This unsharp filter is identical to the example described in Section 4.2, except that it consumes and produces color images, via the initial function gray(x, y) which converts from RGB to grayscale, and the functions ratio(x, y) and unsharp(x, y, c) which apply the sharpening to the color image at the end.

We will continue to use it as a running example throughout this chapter, as it demonstrates many important features of our system. The code first computes a blurred gray scale version of the input image using a chain of three functions (gray, blury, and blurx), and then amplifies the input based on the difference between the original image and the blurred image.

The hardware schedule begins on line 14. unsharp.tile is a standard Halide operation, which breaks an ordinary row-major traversal (defined by vars x and y) into a blocked computation over tiles (here, 256 × 256 pixels). The variables xi and yi represent the inner loops of the blocked computation that work pixel by pixel, while x and y then become the outer loops for iterating over blocks.

With the image now broken into constant-sized pieces, we can apply hardware acceleration. Our first new primitive is f.accelerate(inputs, innerVar, blockVar), which defines both the scope and the interface of the accelerator and the granularity of the accelerator task. The first argument, inputs, specifies a list of Funcs for which data will be streamed in. The accelerator will use these inputs to compute all intermediate Funcs to produce the result f. In this example, this is the sequence of computation through gray, blury, blurx, sharpen, and ratio that produces unsharp from in (Figure 5.1b).

The block loop variable blockVar defines the granularity of the computation: the hardware will compute an entire tile of the size that blockVar counts—in this case, 256×256 pixels. The inner loop variable innerVar controls the throughput: innerVar will increment each cycle, in this case producing one pixel each time. To create higher-throughput hardware, we could use Halide's split primitive to split the innerVar loop into two, and accelerate with the outer one as the hardware stride size.

Our second new primitive is src.fifo_depth(dest, n). It specifies a FIFO buffer with a depth of n, instantiated for the direct edge from function src and function dest. The primi-

```
Func unsharp(Func in) {
  Func gray, blurx, blury, sharpen, ratio, unsharp;
  Var x, y, c, xi, yi;

  // The algorithm
  gray(x, y) = 0.3*in(0, x, y) + 0.6*in(1, x, y) + 0.1*in(2, x, y);
  blury(x, y) = (gray(x, y-1) + gray(x, y) + gray(x, y+1)) / 3;
  blurx(x, y) = (blury(x-1, y) + blury(x, y) + blury(x+1, y)) / 3;
  sharpen(x, y) = gray(x, y) + 0.5 * (gray(x, y) - blurx(x, y));
  ratio(x, y) = sharpen(x, y) / gray(x, y);
  unsharp(c, x, y) = ratio(x, y) * input(c, x, y);

  // The schedule
  unsharp.tile(x, y, xi, yi, 256, 256).unroll(c)
          .accelerate({in}, xi, x)
          .parallel(y).parallel(x);
  in.fifo_depth(unsharp, 512);
  gray.linebuffer().fifo_depth(ratio, 8);
  blury.linebuffer();
  ratio.linebuffer();

  return unsharp;
}
```

(a) Algorithm and schedule code for the **unsharp** function.

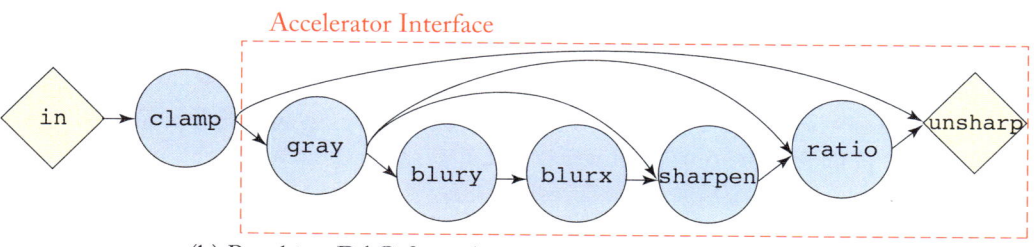

(b) Resulting DAG from the **unsharp** code above.

Figure 5.1: Source code for **unsharp** and the DAG of kernels it produces. The `accelerate` primitive defines the accelerator scope, from *in* to *unsharp*.

tive helps to balance the latencies of different paths between two vertices in a DAG. Unbalanced paths in the DAG will cause pipeline stalls, which results in substantial performance degradation or even deadlock. In the unsharp example, both `in` and `gray` are consumed by multiple functions. Without a FIFO from `in` to `unsharp`, for example, the pipeline stalls after one pixel of `in` is fetched because the channel between `in` and `unsharp` is full while there is no value of `ratio` available for stage `unsharp` to proceed with its computation. In order to eliminate the pipeline stalls, we add a FIFO buffer of depth 512 along the path from `in` to `unsharp` that matches the latency with the other paths. The optimal FIFO depths in the DAG can be solved automatically as an integer linear programming problem as described in Chapter 4, so we can eventually automate this decision, but for now we specify and tune it by hand.[1]

`f.linebuffer()`, our syntactic sugar (not a new primitive) for a combination of existing Halide primitives, is designed to instantiate a line buffer for function `f`.[2] Without these primitives, functions would be fused directly into other downstream functions, potentially causing recomputation when their values were reused. In the unsharp example, by default, function `gray` will be fused into function `sharpen`, `ratio`, and `unsharp`, causing each value of `gray` to be evaluated three times. Instead, we use the `linebuffer` primitive to buffer the value and avoid the recomputation, as shown on line 18 in Figure 5.1a. Therefore, the `linebuffer` primitive helps explore the tradeoff between storage and computation resources available in the hardware.

The existing Halide primitive `f.unroll(var, factor)` is overloaded in the hardware context for specifying variable rate pipeline stages and exploring the space-time tradeoff. For a CPU target, `unroll` is used to eliminate short loops and enable optimizations on cross-iteration sharing of data. However, in terms of hardware, since the HLS tool schedules the resource for one loop iteration, having a larger loop body through unrolling also increases the parallelism of the datapath. It effectively duplicates the compute units in the pipeline, potentially scaling up the throughput. In the example, `unroll` on `unsharp` causes three multipliers to be instantiated for computing three color channels simultaneously, which scales the throughput of the pipeline from 1/3 pixel/cycle to one pixel/cycle.

All other existing Halide primitives (e.g., `tile`, `vectorize`, `parallel`) remain unchanged for the portion of the program mapped to software, where Halide already provides state-of-the-art performance on ARM and x86 CPUs. In our example, the `parallel` primitives on line 16 schedule multiple tile-processing tasks concurrently onto multiple CPU cores and multiple FPGA pipelines.

---

[1] Another issue with the automatic solving for the optimal FIFO depths is that some latencies of hardware blocks are unknown before HLS compilation. In the future, if HLS tool provides API for querying these latencies, the `fifo_depth` primitive can be opted out with the automatic solving implemented in the DSL compiler.

[2] Halide's native analog of the line buffer, the *sliding window* pattern, is achieved by specifying different compute and storage levels with `compute_at` and `store_at` primitives, and letting the compiler apply a *storage folding* optimization. We overload the semantics of this same pattern when it is used in an accelerated portion. As `accelerate` already defines the compute and storage levels, we add the `linebuffer` sugar, which needs no additional arguments.

5.3   COMPILER IMPLEMENTATION

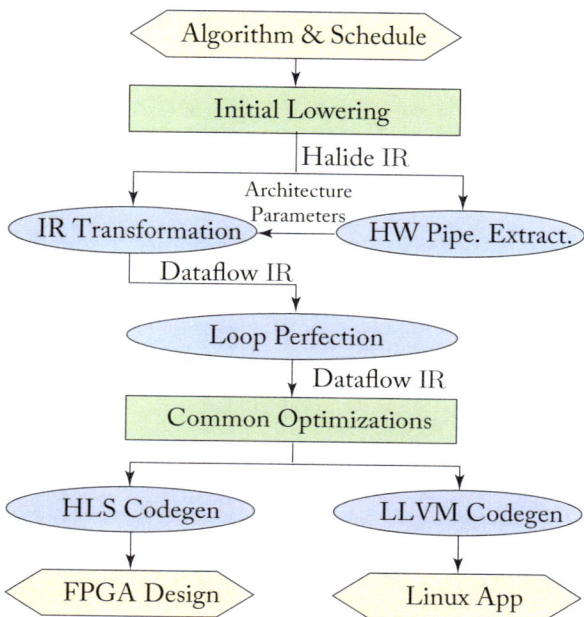

Figure 5.2: Compilation flow. Blue circles are new, green rectangles are unchanged/existing Halide compilation passes. Yellow hexagons are inputs and outputs of the compiler.

Figure 5.2 describes our compiler design. The inputs to our system are an application's *algorithms* and *schedules* written in Halide. An analysis pass extracts parameters for the architecture template. A transformation pass rewrites the hardware parts of the Halide original imperative IR into a dataflow style. After the new loop perfection optimization and some common scalar optimization passes, like constant propagation, common sub-expression elimination, and so on, the final IR is passed to an HLS code generator and an LLVM code generator, which produce the hardware designs and the software programs, respectively.

5.3.1   ARCHITECTURE PARAMETER EXTRACTION

Our system generates specialized hardware accelerators by instantiating architectural templates from a scheduled program. The architectural template is similar to the line-buffered pipeline model that Darkroom uses, with extensions to support a wider range of algorithm and performance targets.

In this template, an accelerator is a DAG whose edges are streams of windows of pixels, or *stencil streams*, and whose nodes are *stencil kernels*. Each kernel is a Halide function scheduled for a line buffer, into which one or more non-line-buffered functions can be fused. A stencil stream is parameterized by the window size, the sliding stride, and the range of the image domain. Note that because an output stencil may serve as input to more than one downstream kernel, the producer kernel must compute a union of the stencils required by all consumers. Figure 5.3 shows some of the architectural parameters for the **unsharp** application of Figure 5.1a.

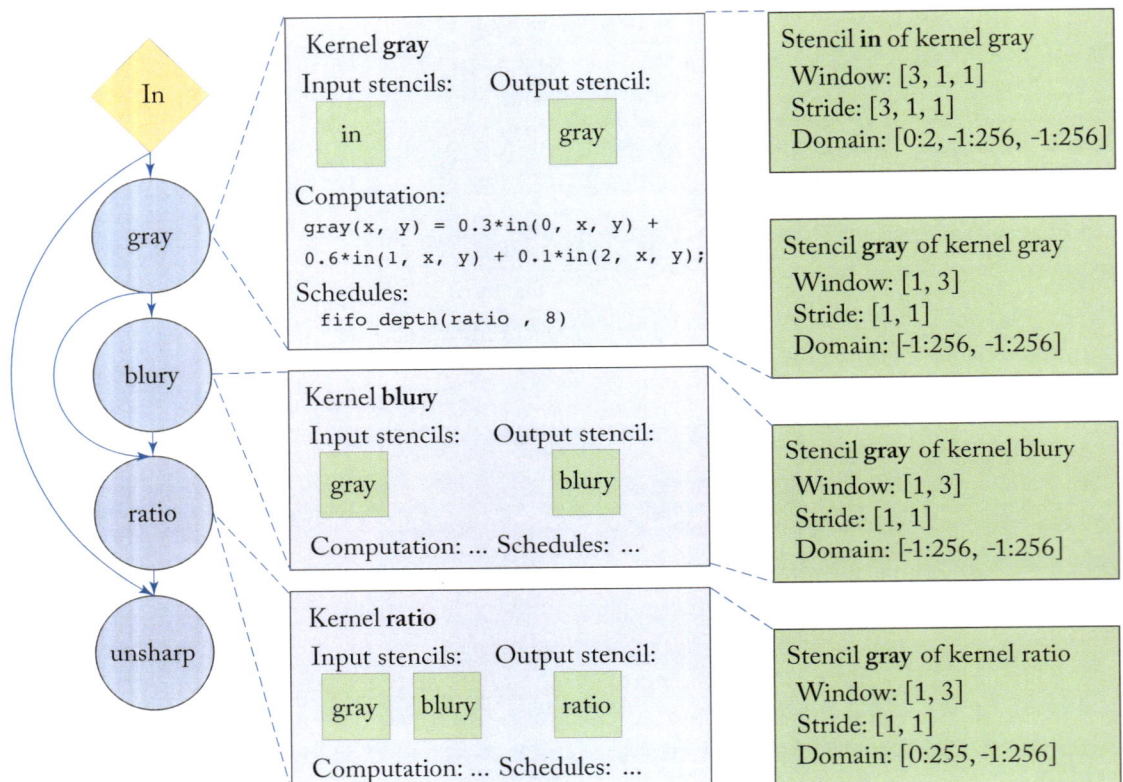

Figure 5.3: Template parameters for *unsharp*. The blue circles are stencil kernels corresponding to line-buffered functions (*blurx* and *sharpen* were fused into *ratio*). Middle and right columns show parameters of selected kernels and stencil streams. The domain of the output stencil stream *gray* is the union of the domains of the stencil *gray* in kernels *blury* and *ratio*.

Since the scope of the pipeline and the line-buffered functions are defined in a high-level language, composing the DAG is straightforward. To extract the parameters of each stencil stream in the template, we apply bounds inference analysis recursively back from the output. This is similar to the original Halide compiler [Ragan-Kelley et al., 2013], except that we now have hardware line buffers between stencil kernels that capture data reuse, so the output stencil size of upstream kernels doesn't cumulatively increase. For example, for a pipeline of three cascaded 3×3 convolution functions, in order to compute one final value, the original bound analysis will infer a 5×5 output of the first function and a 3×3 output of the second function for the worst case (i.e., at image boundaries), while our new analysis is aware of the line buffers that are specialized to handle the boundary conditions and gives 1×1 output bounds for all functions.

5.3.2 IR TRANSFORMATION

Given a schedule, the Halide compiler generates an architecture-independent imperative representation of the algorithm, with loop nests and storage allocations injected for each function. Figure 5.4a shows part of the Halide loop IR for function *gray*. Because the schedule specifies a sliding window order for computing *gray*, the storage for *gray* is allocated outside the loop nest that iterates over the 256×256 block of *unsharp*, while the computation of *gray*, along with other functions, is interleaved inside the loop nest.

To map this scheduled loop IR into the line buffer pipeline shown in Figure 5.3, we further lower it into our own dataflow IR, using the previously extracted architecture template parameters. First, additional storage for input and output stencils are allocated locally to a kernel computation, and references to the original storage are replaced with references to stencils. Then, the original storage allocation is replaced with declarations of stencil streams, and stream operations (pop and push) are inserted before and after the kernel computation. Finally, loops iterating over the image domain are inserted for each compute stage, along with calls to the linebuffer and dispatch IR primitives if a line buffer or a stream dispatcher is needed.

Figure 5.4b shows the IR of kernel *gray* after the transformation. For each scanning step (here, each scan_x loop iteration), the inner loop nest computes a 1×1 *gray* stencil and pushes it to gray_step_stream, which is later line-buffered and dispatched to the kernels *blury* and *ratio*. The unit length loops x and y will be eliminated in a later simplification pass.

After the IR transformation, the dataflow IR presents an untimed, bit-accurate representation of the pipeline, with different stages explicitly separated by data streams. For example, the **unsharp** pipeline is composed of four kernel stages, two line buffers, and two stream dispatchers, as shown in Figure 5.5. Many HLS tools, such as Vivado HLS [Xilinx, 2016] and Catapult HLS [Mentor Graphics, 2016], can infer coarse-grain pipelined designs from such code structure. However, the throughput of stages producing less than 1 pixel/cycle is not optimal due to limits on the automatic loop pipelining of the current HLS tool. We address this problem next.

5.3.3 LOOP PERFECTION OPTIMIZATION

Figure 5.6a shows the IR of a five-point convolution kernel scheduled at the rate of 1/5 pixel/cycle. Loop pipelining in Vivado HLS only applies to a *perfect loop nest* such that only the innermost loop has operations. Here, loops scan_x and d do not form a perfect loop nest due to the instructions on line 2, 3, and 6, and thus the pop, initialization, accumulation, and push operations run sequentially and cannot be pipelined in the generated hardware. Moreover, it is expensive in hardware to enter or exit a loop due to the pipeline flushing overhead. In this case, the pipeline for the innermost loop (loop d) needs to be flushed every five iterations.

```
1   alloc gray [258, 258]
2   for (unsharp.y, 0, 256)
3     for (unsharp.x, 0, 256)
4       // compute function "in" here...
5       for (y, gray.y.loop_min, gray.y.loop_extent)
6         for (x, gray.x.loop_min, gray.x.loop_extent)
7           gray(x,y) = 0.3*in(0,x,y) + 0.6*in(1,x,y) + 0.1*in(2,x,y)
8       // compute function "blury" here...
```

(a) Halide IR of function *gray*.

```
1   // kernel "in" here...
2   def_stream gray_stream [1, 3]
3   def_stream gray_step_stream [1, 1]
4   for (scan_y, 0, 258)
5     for (scan_x, 0, 258)
6       alloc in [3, 1, 1]
7       alloc gray [1, 1]
8       pop (in, in_stream)
9       for (y, 0, 1)
10        for (x, 0, 1)
11          gray(x,y) = 0.3*in(0,x,y) + 0.6*in(1,x,y) + 0.1*in(2,x,y)
12      push (gray, gray_step_stream)
13  linebuffer(gray_step_stream, gray_stream, [258, 258])
14  dispatch(gray_stream, [-1:256, -1:256],
15          "blury", 1, [-1:256, -1:256],
16          "ratio", 8, [0:255, -1:256])
17  // kernel "blury" here...
```

(b) Dataflow IR of kernel *gray*.

Figure 5.4: IR transformation allocates additional storage for local stencils and separates pipeline stages with data streams.

In order to create a fully pipelined convolution stage with 1/5 pixel/cycle throughput, the content in loop scan_x must be pushed into the innermost loop, as shown in Figure 5.6b. The perfect loop nest after the transformation not only creates a longer hardware pipeline containing all the operations, but also eliminates the pipeline flushing overhead when entering an intermediate loop level. In hardware, the "if" statements are usually implemented using multiplexers that select the results from true branches. Nevertheless, for I/O operations, such as the push and pop operations in the example, the multiplexers select the values of the control signals of the I/O ports, so the real I/O operations only happen when the predicates are true.

We apply this restructuring automatically in the accelerated region of IR through a recursive descent algorithm. Note that loop perfection is an inverse operation of loop peeling [Wolfe,

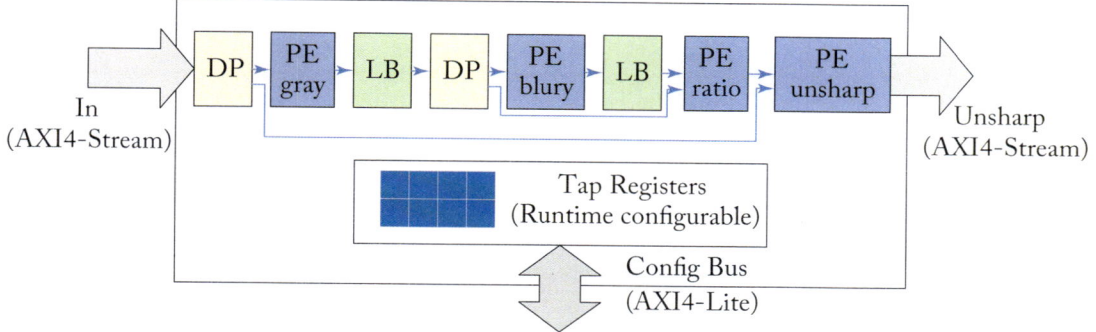

Figure 5.5: Pipeline implementation for *unsharp*. Processing elements *PE* implement kernel DAGs. Line buffers *LB* capture the data reuse in the stencil pattern. Stream dispatchers *DP* fork a stencil stream to multiple consumers.

```
for (scan_x, 0, 256)
    pop (in, in_stream)
    out(0) = 0
    for (d, 0, 5)
      out(0) += mask(d)*in(d)
    push (out, out_stream)
```

(a) Original IR .

```
for (scan_x, 0, 256)
    for (d, 0, 5)
      if (d == 0)
        pop (in, in_stream)
        out(0) = 0
      out(0) += mask(d)*in(d)
      if (d == 4)
        push (out, out_stream)
```

(b) IR after the optimization.

Figure 5.6: Loop perfection optimization creates a larger perfect loop nest by pushing operations from the outer loop body into the innermost loop.

1992], which moves predicates out of the innermost loop to improve performance on traditional processors, as it reduces the number of dynamic instructions.

5.3.4   CODE GENERATION

After some common optimizations, the final IR of the pipeline is passed to two different code generator backends, HLS and LLVM. The HLS code generator translates the hardware accelerator portions of IR into HLS-synthesizable C code, and the rest of the IR is translated into a C++ testbench wrapper. The code generator inserts HLS directives (pragmas) automatically to assist the HLS compiler in applying loop pipelining and array partitioning. To simplify the code generator, we built an HLS-synthesizable C++ template library implementing an abstract line buffer interface:

```
template<int IMG_SIZE_0, int IN_SIZE_0, int OUT_SIZE_0, int IMG_SIZE_1, int
    IN_SIZE_1, int OUT_SIZE_1, ...>
void linebuffer(stream<IN_SIZE_0, IN_SIZE_1, ...> &in,
                stream<OUT_SIZE_0, OUT_SIZE_1, ...> &out);
```

We designed the multidimensional (up to 4D) line buffer templates hierarchically, as shown in Figure 5.7.

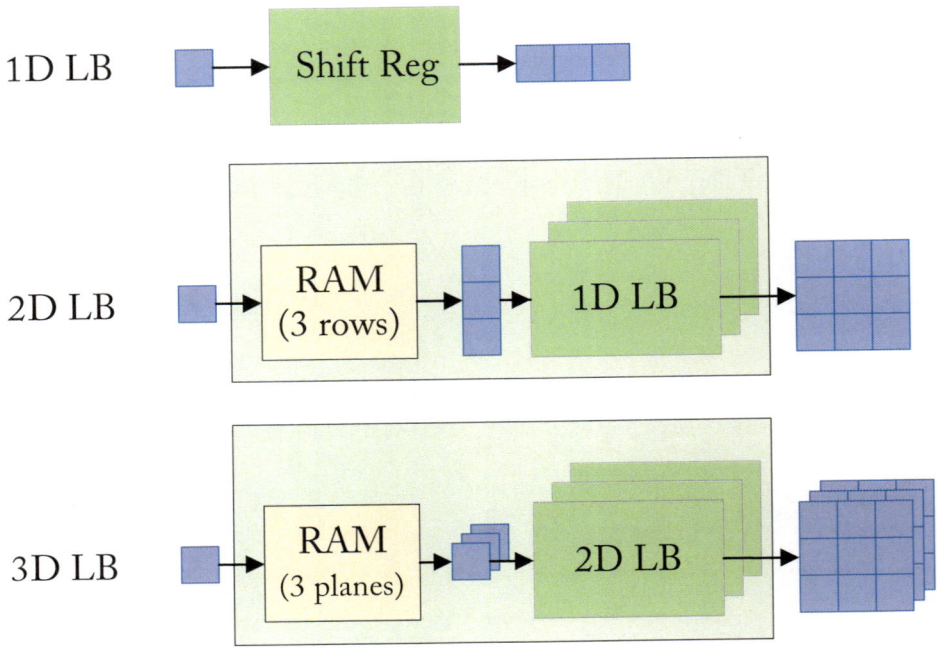

Figure 5.7: Multidimensional line buffer. 1D LB uses shift register. 2D LB uses RAM to buffer rows of pixels, pushes a column to 1D LBs and outputs 2D stencils. 3D LB instantiates RAMs and 2D LBs similarly.

As a further design optimization, the compiler can also statically evaluate constant functions (e.g., lookup tables), and generate code that later synthesizes to ROMs.

Generating machine code for the CPU portion of the software program is left to the LLVM compiler infrastructure. We largely use the existing Halide ARM backend, which includes a highly optimized ARM NEON SIMD vectorizer, thread-pool based parallel runtime, and so on. The final IR describes a complete pipeline with both CPU and accelerator, from which the code generator emits platform-specific device driver calls to access the hardware when it visits the boundary of the hardware pipeline during IR traversal. Moreover, the tool recognizes all the data buffers accessed by any hardware pipeline and thus can emit special allocation routines for these buffers, inserting data transfers where required by the platform setup.

5.4   IMPLEMENTATION AND EVALUATION

As with Darkroom, we implemented the Halide system on the Xilinx Zynq platform, in this case a ZC702 evaluation board holding a Zynq XC7Z020 SoC. This has a somewhat smaller FPGA fabric compared to the XC7Z045 used in the previous chapter but is more widely available, most notably on the low-cost ZedBoard platform. The low power of this board (total system power of 2 W) makes it an ideal platform for running many image-processing applications in mobile and battery-powered environments. We run Linux (kernel version 4.0, built from Xilinx Open Source Linux 2015.4) on the ARM cores with a filesystem built from Ubuntu Base, giving us the many conveniences of a full operating system (e.g., a file system, networking, and core utilities), and also realistically modeling the challenges of integrating an accelerator into a larger heterogeneous system.

5.4.1   PROGRAMMABILITY AND EFFICIENCY

We implemented six individual applications (a 3×3 **gaussian**, **corner**, **unsharp**, **stereo**, **bilateral grid**, and a version of **campipe** from the Frankencamera project [Adams et al., 2010]), plus an application that combines **campipe** and **unsharp** in Halide, and generated the hardware and software for Zynq using our compiler.

Table 5.1 lists the specifications of generated accelerators for these applications. In the evaluation, we used the optimal schedule we found for each application, including the best workload partitioning between CPU and FPGA. Since the Zynq ARM cores are relatively weak as compared to the FPGA fabric, for most of the applications, the CPU does not implement any computation kernels but simply controls tiling, i.e., calculating the coordinates of each image tile and scheduling the accelerator for processing tiles. In **stereo** the CPU also computes padding using a repeat edge condition, and in **bilateral grid** it shuffles data into an 8×8 grid. We run 8-megapixel images through each application, but the BRAM usage for internal buffering in each accelerator is kept low thanks to the image tiling.

For comparison to a CPU+GPU platform, we use an NVIDIA Jetson TK1 board with JetPack 2.0. The Tegra K1 SoC is fabricated in the same 28 nm technology as the Zynq SoC. We evaluate both ARM and CUDA implementations on TK1 for each application using the same Halide algorithm but different schedules, optimized by Halide experts. For **stereo**, where Halide-generated CUDA performs poorly, we compared to the Tegra-optimized OpenCV CUDA kernel shipped by NVIDIA with JetPack.

To measure power on the ZC702, we pull statistics from the onboard power controllers, which report the power of each subsystem including FPGA, DRAM, and CPU. For the TK1 board, we measure the current on the 12V DC supply. To derive the energy efficiency for TK1, we subtract the board idle power (about 2 Watts) to exclude the board's uncore components, which gives the TK1 an advantage as it also excludes the SoC and DRAM static power. We use `gettimeofday` to measure software program execution time and report the average over 20 runs.

Table 5.1: Specifications of generated accelerators for the evaluated applications. The resource utilization is reported as percentage of LUT, FF, BRAM, and DSP used by the given application on the Zynq XC7Z020.

Application	Rate (pix/cyc)	Read BW (MB/s)	Resource (%)			
			LUT	FF	BRAM	DSP
gaussian	2	200	9	5	3	33
corner	2	284	22	19	7	30
unsharp	1	375	7	5	7	31
stereo	0.25	62	55	32	5	0
bilateral grid	1	182	23	20	14	13
campipe	2	569	9	6	5	5
campipe + unsharp	1	250	14	10	11	34

For the CUDA target, we exclude the data transfer time between CPU and GPU (again giving the TK1 an advantage).

Figures 5.8 and 5.9 show the throughput and energy efficiency of the Zynq implementation versus the TK1's ARM and GPU cores, respectively. On average, applications on Zynq achieve 2.6× and 1.9× higher throughput than CPUs and GPU, respectively, and 14.2× and 6.1× better energy efficiency. **Corner** achieves the most energy reduction of 38× and 12×, as well as the highest throughput speedups of 6× and 3.5×, compared to the TK1 CPUs and GPU. The energy efficiency is achieved by high locality and data reuse exploited in the line-buffered accelerator pipeline and the greatly reduced memory requests as compared to programmable cores. In addition, the applications mostly use low-precision fixed-point arithmetic, which can be efficiently implemented using the LUTs and DSPs on an FPGA fabric.

All Zynq-based applications except **campipe** achieve the peak throughput of the accelerator in Verilog simulation. The **campipe** accelerator requires much higher read bandwidth (568 MB/s), and thus suffers the most from bandwidth problems caused by reading data that misses in the L2 cache (see Section 6.4.5).

As long as the application fits in the FPGA fabric, the throughput of the Zynq implementation is generally bound by memory bandwidth and clock frequency. Therefore, speedups compared to the ARM CPUs or GPU on TK1 are proportional to the number of operations accelerated on the FPGA (approximately proportional to the number of LUTs and DSPs used). For this reason, **corner**, **stereo**, and **campipe+unsharp** get the most speedup. In **bilateral grid**, the accelerator also uses a lot of LUTs, but most of them implement control logic and multiplexers for building histograms and data-gathering for interpolation, which do relatively little real computation. Moreover, parallelism in the kernels of the **bilateral grid** is limited by data depen-

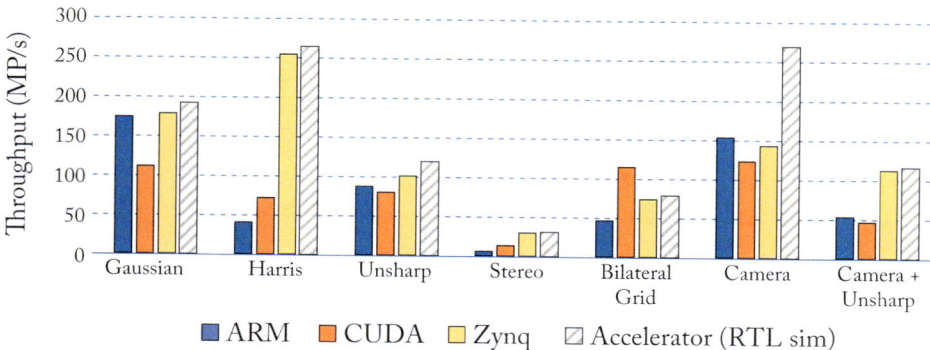

Figure 5.8: Throughput comparison of the four TK1 ARM cores, CUDA GPU on TK1, our Zynq platform implementation, and an ideal accelerator. The accelerator RTL simulation value represents the theoretical throughput for the accelerator if it were never bottlenecked by other parts of the system.

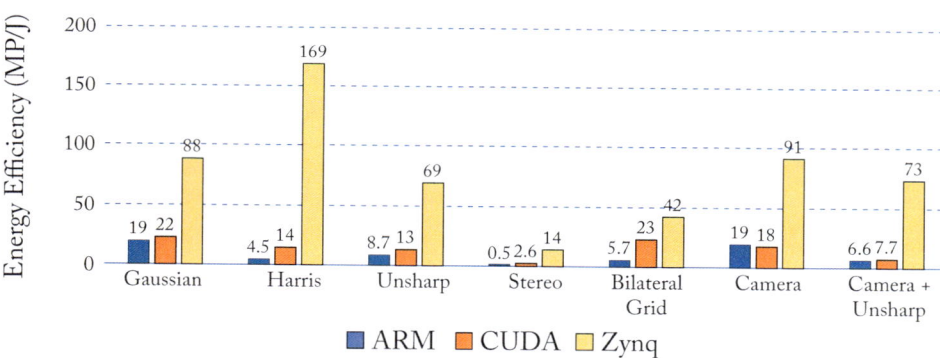

Figure 5.9: Energy efficiency comparison of the four TK1 ARM cores, CUDA GPU on TK1, and the Zynq platform.

dencies, making it more favorable to execute on high frequency and high memory bandwidth processors (i.e., the Tegra CPUs and GPU).

An important takeaway is that acceleration using the FPGA becomes more effective as the image-processing pipeline grows deeper, which matches the trend of new applications from computational photography and computer vision. For example, on CPU or GPU, the execution time of the **campipe+unsharp** combination is the sum of the execution times of the two individual applications. However, pipelining two applications onto the FPGA fabric simultaneously doesn't increase the required memory bandwidth or slow the clock frequency. Therefore, the throughput for any composition of pipelines that fit on the FPGA is bounded only by the slowest app in the chain (in this case, **unsharp**, which is around 110 megapixels/second).

Figure 5.10 plots the energy cost per operation and the compute intensity for different applications on Zynq. The FPGA energy ranges from 2.6 pJ/op to 35 pJ/op depending on the types of operations, while the CPU and DRAM energy scales strongly with the compute intensity. **Stereo** achieves the lowest energy, 5.8 pJ/op, with the highest compute intensity of 4220 op/byte. On the other hand, the 26 op/byte application, **bilateral grid**, consumes 311 pJ per operation, 186 pJ of which is spent on DRAM accesses.

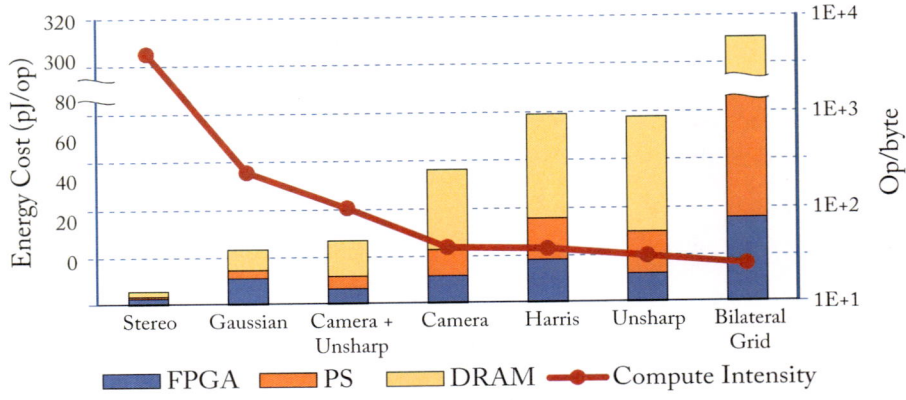

Figure 5.10: Energy cost breakdown and compute intensity in term of operations per byte access to DRAM for different applications on Zynq. The energy cost reduces with more operations accelerated onto FPGA.

5.4.2   QUALITY OF HARDWARE GENERATION

To evaluate the quality of generated FPGA implementations, we develop two applications in Halide, a 5×5 2D convolution (**convolution**) and **corner**, which are also available in both the Xilinx HLS video library [Xilinx, 2016] and in the example repository of the open-source HIPAcc-Vivado compiler [Membarth and Reiche, 2016].

Halide is able to generate hardware with similar quality to that of these other two frameworks. While it is hard or even impossible to code all the applications in Chapter 3 using the other frameworks, **convolution** (a single stage kernel) and **corner** (a pipeline of many kernels), suffice to showcase the hardware building blocks and their compositions, respectively.

Generated FPGA designs for the three frameworks achieve similar peak frequencies of around 180 MHz for **convolution** and 150 MHz for **corner**. The **corner** pipelines operate slower because they are bounded by a floating-point-to-integer conversion. Figure 5.11 shows the FPGA resource utilization of our generated designs vs. library and HIPAcc-Vivado designs on a 1080 p image. We generated designs at different pixel rates using the same Halide algorithm code but different unrolling schedules, while the library and HIPAcc only provide designs at a single pixel per cycle rate.[3]

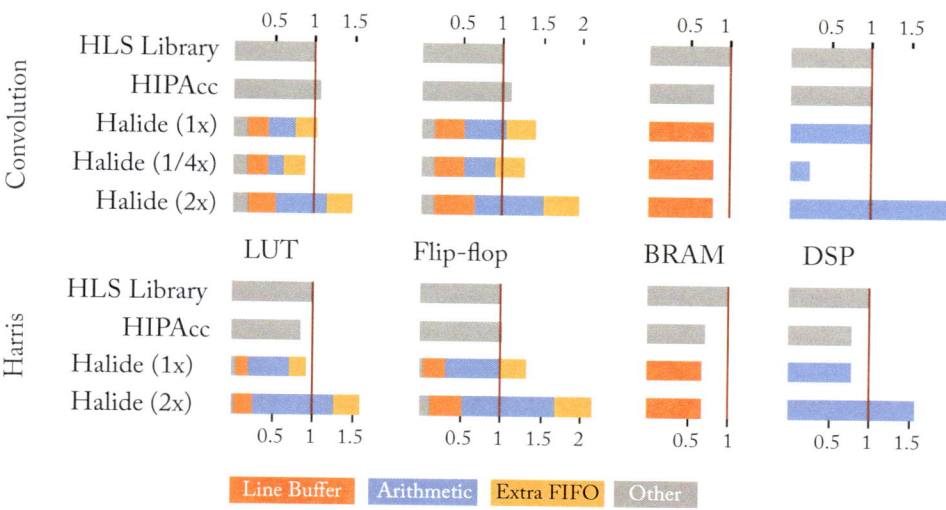

Figure 5.11: Resource utilization (area) comparison of **convolution** (top) and **corner** (bottom) in our system at various pixel rates vs. HLS library and HIPAcc implementations. Our system's designs can target multiple different throughput, while HIPAcc and library kernels are only optimized at a 1 pixel/cycle rate.

Halide's line buffer components use less BRAM for two reasons. First, the line buffer instance from the library is not optimal in terms of storage usage: in 5 × 5 convolution, the library design buffers five rows while only four are required, so both our and HIPAcc's designs start with 20% less BRAM. Second, the library and HIPAcc both instantiate per-kernel line buffers for each input stencil, while we place a line buffer for each *output* stencil stream, which can be shared among kernels consuming the same stream. In **corner**, our design instantiates fewer line buffer instances thanks to this buffer sharing, for an extra 6% BRAM savings as compared to HIPAcc.

In **corner**, our design and HIPAcc use fewer DSPs and LUTs because the code generation (meta-programming) approach is more flexible and does better constant propagation and simplification as compared to the C++ template solution used by the library. Our unit-rate **convolution** design has fewer LUTs because the library solution creates four separate streams for the RGBA channels, whereas we package RGBA as a single structure and pass it through a wider stream along the pipeline, which simplifies the control logic for managing data streams.

---

[3] The recent HIPAcc Altera-OpenCL backend [Özkan et al., 2016] can generate >1 pixel/cycle pipelines through kernel vectorization, but cannot generate <1 pixel/cycle designs. We did not evaluate this system as it is not opensource.

Multirate designs (higher than 1 pixel/cycle) need higher compute throughput and buffer bandwidth. Fortunately, the BRAM banks of the line buffers provide more than enough bandwidth for 1080p images. Therefore, the multirate designs use more LUTs and DSPs for the arithmetic datapath, while the line buffer resources do not change significantly.

One source of overhead in our design comes from unnecessary FIFOs inserted between pipeline stages. We use `hls::stream` objects to connect different stages in the generated HLS code, and the current HLS compiler creates a hardware FIFO for each stream object. However, in our design, stages can be directly connected through a handshake interface (e.g., AXI4-stream) because the latencies in the pipeline are already balanced. The extra FIFO adds 30% FF and 20% LUT overheads. Future optimizations in the HLS compilation could eliminate these unnecessary FIFOs.

Table 5.2 summarizes the code length of **convolution** and **corner** using different systems. Moving from less to more domain specific, Halide and HIPAcc DSLs are orders of magnitude more compact than both HLS C library and our generated HLS code. Because Halide uses functional representation, its application code is 2× shorter than HIPAcc's. While code length is an imperfect indicator of design productivity, the vastly reduced size of HIPAcc and Halide code demonstrates the ability of these high-level languages to concisely capture a design.

Table 5.2: Lines of code (LoC) for **convolution** and **corner** in HLS library, HIPAcc, Halide, and generated HLS code from Halide. HLS Lib excludes basic data structure code; HIPAcc counts are DSL parameter+algorithm [Membarth et al., 2016]; Halide counts are algorithm+schedule; Generated HLS counts exclude 900 LoC in the line buffer template library.

	HLS Lib	HIPAcc	Halide	Generated HLS
Conv2D	209	5 + 5	2 + 2	885
Harris	520	47 + 26	23 + 11	619

5.5   CONCLUSION

The Halide image processing DSL enables developers to quickly create and optimize new image-processing applications, and in this chapter, we have demonstrated extensions to Halide that allow it to target FPGAs. Particularly useful is the fact that the Halide code is used to generate both the design of the accelerator and the software that communicates with the hardware. Like Darkroom, the high-level constructs in Halide permit us to map the algorithm to hardware using very efficient design patterns so that the resulting designs are competitive with other approaches while providing vastly improved design productivity vs. handwritten HDL or C-based high-level synthesis.

In the following chapter, we move to the software side of the system and look at how an accelerator is integrated with the rest of the system.

CHAPTER 6

Interfacing with Specialized Hardware

Up to this point, we have focused on the problem of creating a "program" for a piece of specialized hardware, whether that is a sequence of instructions, register configuration, FPGA bitstream, or something else. But as we described in the introduction, this is only half the battle.

In this chapter, we move a layer up the stack and consider another question: Once we have a piece of specialized hardware configured with a program, how do we connect it to the rest of our system and integrate it into our application? We'll first look at the common interfaces to specialized hardware and then consider some of the specific things that make this particularly hard. Then we'll explore a number of solutions to these problems and show how domain-specific languages and other techniques can be combined to simplify this problem.

6.1 COMMON INTERFACES

Although accelerators are numerous and varied, there are only a handful of ways that they can be connected to systems, ranging from tightly to loosely coupled [Caşcaval et al., 2010]:

- On the CPU datapath, controlled by special instructions. Examples include SIMD and cryptography accelerators.

- Memory-mapped, controlled by a set of registers accessible to the main processor via the memory bus. The device may also be able to perform direct memory access (DMA) transactions to read or write memory without the intervention of the host CPU.

- On a system bus (e.g., PCI Express). The operating system provides facilities to map device memory into the host process memory space, much like the memory-mapped case. The systems bus typically supports DMA transfers as well.

- Across a network connection, controlled by command packets set via the network.

We will focus on the image-processing accelerators we have described in the previous chapters, which are typically connected to the processor via memory mapping on the data bus. They also have access to memory via DMA. Thus, for the accelerators described in the previous two chapters, all operations happen through one of two distinct interfaces: setting configuration via memory-mapped registers and streaming blocks of image data via DMA.

6.2   THE CHALLENGE OF INTERFACES

Despite the relative simplicity of the interfaces themselves, constructing the bridge between the hardware and the application remains a formidable task. Interfacing with specialized hardware shares many of the same complications as multithreading, plus a few more.

1. An accelerator that does any significant amount of work per invocation (i.e., anything not on the datapath) will run asynchronously with respect to the host CPU; this implies the need for some kind of resynchronization when the work completes.

2. The hardware must continually be fed new tasks; this may require constant tending by the host.

3. Timing matters. Sometimes registers need to be set at precise times, or actions performed in a particular sequence with specific delays between them.

Then there are the problems associated with memory:

4. Coherency needs to be carefully managed, and the accelerator may have special memory requirements. In some systems, the accelerator can be connected into the main cache hierarchy and coherency is handled automatically. For example, on ARM chips, the Accelerator Coherency Port provides a bus for accelerator memory accesses that goes through the cache hierarchy. In other cases, the cache system isn't able to help and the software must handle cache flushing and invalidation directly.

5. If the accelerator cannot handle paged memory (i.e., scatter-gather memory), then the software must provide special contiguous buffers. Likewise, some accelerators have a smaller address space than the host platform and can only access a fraction of system memory. In both cases, the software must manage the transfer between regular user-space memory and the special accelerator-accessible buffers. There is a similar memory-transfer issue for accelerators on a system bus that have their own memory, such as a GPU attached to a PCIe bus.

Of course, a well-written kernel driver will hide most of this ugliness, and a well-written API layered on top will hide nearly everything. But it isn't sufficient to simply say that "the driver will take care of it." Writing kernel drivers is difficult; debugging them even more so. User-space code doesn't have the memory-allocation and interrupt-handling capabilities that some hardware devices need, and there is no other way to access these facilities. Our challenge here is to make it easier to create these drivers and APIs.

6.3   SOLUTIONS TO THE INTERFACE PROBLEM

6.3.1   COMPILER SUPPORT

One of the cleanest ways to integrate support for an accelerator is to support it in the compiler. This is particularly effective where the accelerator is built into the CPU datapath, as is the case for vector units, math coprocessors, or cryptography accelerators. In all these cases, the accelerators are invoked with special CPU instructions and operate synchronously, with relatively small latency.

In simple instances it is sometimes possible for the compiler to infer that that the hardware can be used to improve performance. For example, to make use of SIMD vector units, gcc and other compilers can automatically replace element-by-element for loops with vectorized versions that do the same thing. In most other cases, however, the accelerator must be explicitly invoked by writing inline assembly code or special instructions that the compiler maps to the hardware.

Another example of integrating accelerator support into the compiler is the customizable Tensilica Xtensa processor.[1] A designer can customize the CPU and attach his or her own specialized accelerators to it by extending the instruction set. The compiler and software development environment are extended more or less automatically.

6.3.2   LIBRARY INTERFACE

For fixed-function hardware that operates asynchronously, the obvious solution is to wrap accelerator access into a library function call, or set of calls. This works well where configuration is simple and a function call encapsulates the parameters that need to be passed in. One such example is an H.264 encoder, where a handful of parameters and the input frame itself are sufficient to control its operation.

A library approach makes integration straightforward, but it doesn't help with creating the library itself. If the library is created by direct specification (e.g., writing kernel driver code in C), then it will have many of the problems we noted in Chapter 2: it is time consuming to create, and probably not portable to other systems. Additionally, the fact that the hardware and software components are created independently is a source of interface bugs.

6.3.3   API PLUS DSL

What should we do for more complex accelerators, whose behavior cannot be controlled with a simple API call? It's clear there is no generic driver for an FPGA, or even a generic driver for an image-processing accelerator. Here we can take a cue from GPUs. Like an FPGA or a generic image processing accelerator, a GPU does not have a single program to execute. Instead, it can run a great many "shader" programs, and the software interface depends on which shader is executing.

---

[1]Tensilica was acquired by Cadence in 2013.

This problem is tackled by OpenGL and Direct3D, which are part of the reason real-time graphics has been a success. Both OpenGL and Direct3D provide a "shader language," plus a user-level API to interact with the GPU and the shader program. The shader language is a high-level domain-specific language: it severely restricts the domain (to inherently parallel computations on vertices or pixels), which allows the shader compiler to produce very high-performance implementations for the GPU. This combination of an expressive API and a high-level language help OpenGL and Direct3D (and now Vulkan) to form a very effective abstraction layer between hardware and software. The hardware can change, but as long as it provides a shader compiler and upholds the API, the application should still run (hopefully faster!). This has been an incredibly successful strategy for graphics, and we can take a similar approach for other heterogeneous systems. Darkroom and Halide can serve as the "shader languages" for image processing on an FPGA; now we need the driver and API.

6.4 DRIVERS FOR DARKROOM AND HALIDE ON FPGA

The driver needs to provide a simple interface for controlling the accelerator while addressing the challenges in Section 6.2. Our implementation targets the Zynq SoC running Linux, but the basic operations apply on other systems, even if the implementation mechanisms are different.

The Unix driver interface is often summarized by the phrase "everything is a file." That is, every device is represented by a file node in /dev, and this can be opened, closed, and read from like other files. Of course, not every possible device operation fits into the read()/write() paradigm, so there are some additional operations to cover these cases.

More concretely, Linux device drivers implement one or more of the following interfaces.

- read and write, which are the common operations on files. The Video4Linux framework, for example, implements read for the user to read a frame from a webcam or other device.

- mmap, which maps the contents of a file into the caller's memory space. This is typically used to create a userspace mapping into the device address space, which allows subsequent user code to read and write to registers or device memory with very little overhead.

- ioctl, which is a sort of catch-all for other operations that drivers need to perform but which do not fit neatly into the other interfaces. For example, a webcam driver might use ioctl to set the resolution or white balance settings.

Since the hardware reads and writes blocks of memory (much like Video4Linux devices), the read/write interface seems like a natural choice. However, read and write operate on userspace buffers, which are paged and scattered all over physical memory. An accelerator can deal with this by walking through a separate table in memory that has been constructed with pointers to all the pages, pulling the data from each page in succession. The Xilinx DMA engines include this "scatter-gather" capability, allowing the rest of the accelerator to be agnostic to where the data came from in memory. However, if the page size is 4 kB, a 1080p 24-bit RGB images requires over 1500 pages, and simply building the data structure and triggering cache flushes for each page consumes an inordinate amount of time on the CPU. Perhaps more importantly, storing images in contiguous memory allows the system to access sub-blocks of the image simply by adjusting the base pointer and the width and height. This is particularly helpful when we start stripping or blocking the image to reduce the buffering requirements of the accelerator.

The second option is `mmap`. By mapping the accelerator registers into user memory space, the user has direct control of the hardware. However, this does basically nothing in terms of abstraction, leaving all of the aforementioned interface challenges unresolved. In many cases, `mmap` can be used as a minimal kernel-mode driver that provides raw access to the hardware while a user-space driver does the heavy lifting. However, not all operations are available in userspace, as we'll describe later.

Instead, we chose to implement the interface primarily with `ioctl`, using custom commands to dispatch and fetch contiguous buffers. An `ioctl` command takes the following form:

```
ioctl(int fd, unsigned long request, void* argp)
```
were the three parameters are:

- `fd`, the file descriptor which has been `opened`;

- `request`, which specifies which command to run. These commands are defined in a header file accompanying the driver; and

- `argp`, which is a pointer to one or more arguments for the command.

This is a raw interface with no safety net, but it provides the most flexibility of the available driver interfaces in Unix. In a sense, each `ioctl` sub-command is like its own syscall, with its particular semantics and behavior.

6.4.1 MEMORY AND COHERENCY

We solve challenge 5 by using the Linux contiguous memory allocator (CMA). A kernel boot parameter (`cma=200MB`) reserves a chunk of memory, and then the Linux `cma_alloc_coherent()` kernel operation transparently makes contiguous allocations from this space.

Because memory operations are independent of the hardware, they are wrapped in a separate driver, with two `ioctl` operations, GET_BUFFER and FREE_BUFFER. The former allocates a buffer and populates a structure with pointers for it; the latter releases the buffer to be used again elsewhere. To access the contents of the buffer, it must be mapped into the program memory space, which is performed with `mmap`. A typical usage is shown in Figure 6.1.

We handle coherency between the CPU cores and FPGA accelerator (problem 4) by using the ARM Accelerator Coherency Port (ACP) integrated into the Zynq chips. Memory transactions through this special AXI port are kept coherent with the CPU cache.

An alternative is to explicitly flush and invalidate the cache. Linux has mechanisms to handle this, integrated as part of the `dma_map_single` and `dma_unmap_single` kernel calls. In

```
1  int cma = open("/dev/cmabuffer0", O_RDWR);
2
3  Buffer buf;
4  buf.width = 256; // Width of the image
5  buf.height = 256; // Height of the image
6  buf.depth = 4; // Bytes per pixel
7  buf.stride = 256; // Pixels between successive lines
8
9  // Request a buffer of the size specified in buf
10 int ok = ioctl(cma, GET_BUFFER, (long unsigned int)&buf);
11
12 // Get a userspace pointer
13 long* data = (long*) mmap(NULL, buf.stride * buf.height * buf.depth,
14                       PROT_WRITE, MAP_SHARED, cma, buf.mmap_offset);
15
16 // Fill the buffer with data, and do things with it
17 // ...
18
19 // Release the userspace pointer (and the corresponding kernel structures)
20 munmap((void*)data, buf.stride * buf.height * buf.depth);
21
22 // Release the buffer
23 ok = ioctl(cma, FREE_BUFFER, (long unsigned int)&buf);
```

Figure 6.1: Example usage of the cmabuffer driver to manage contiguous memory buffers from userspace. Error handling code is omitted for brevity, but proper usage would check for failures after the open, ioctl, and mmap calls.

our experiments, we found that these calls took approximately as much time as actually transferring the data, so we elected to use the ACP instead. On other hardware with faster cache flush and invalidate operations, or without a coherent memory bus, the dma_[un]map_single operations might be more appropriate.

6.4.2   RUNNING THE HARDWARE

A second driver handles the interface with the accelerator itself. The essential ioctl command exposed by this driver is PROCESS_IMAGE:

```
1  int id = ioctl(hwacc, PROCESS_IMAGE, (long unsigned int)bufs);
```

Here, bufs is a pointer to an array of Buffer structs from the CMA allocation driver. The array contains at least an input and output image, but depending on the algorithm, may contain multiple input images and/or small images holding "tap values" such as convolution weights.

The driver maintains a queue of images to be processed, and each call to PEND_PROCESSED drops another set of buffers into that queue. Each time the driver receives a completion interrupt, it pops the next request (if any) off the queue, configures the DMA engines with the buffer addresses, and sets the hardware to run. Having an input queue greatly relieves the timing burden from the user. As long as there is at least one item in the queue, the driver will keep feeding the hardware, and it will run at full occupancy. Additionally, this removes the burden of setting configuration registers in between processing runs. The configuration is bundled with the input and output buffers, and the driver applies the values immediately before kicking off the hardware.

Because the accelerator will be busy for hundreds or thousands of microseconds at a time, it is not appropriate for this call to block (problem 1). Instead, it pushes the set of buffers onto the queue and immediately returns a unique "tag" value. Later, the user code can make another ioctl call, PEND_PROCESSED, passing in the same tag. This call will block until the operation is complete, indicating that the buffer set is free to be processed by the CPU again. If the operation has already finished, the call simply pops the buffer off the output queue and returns immediately. The overall operation of the hardware driver is shown in Figure 6.2.

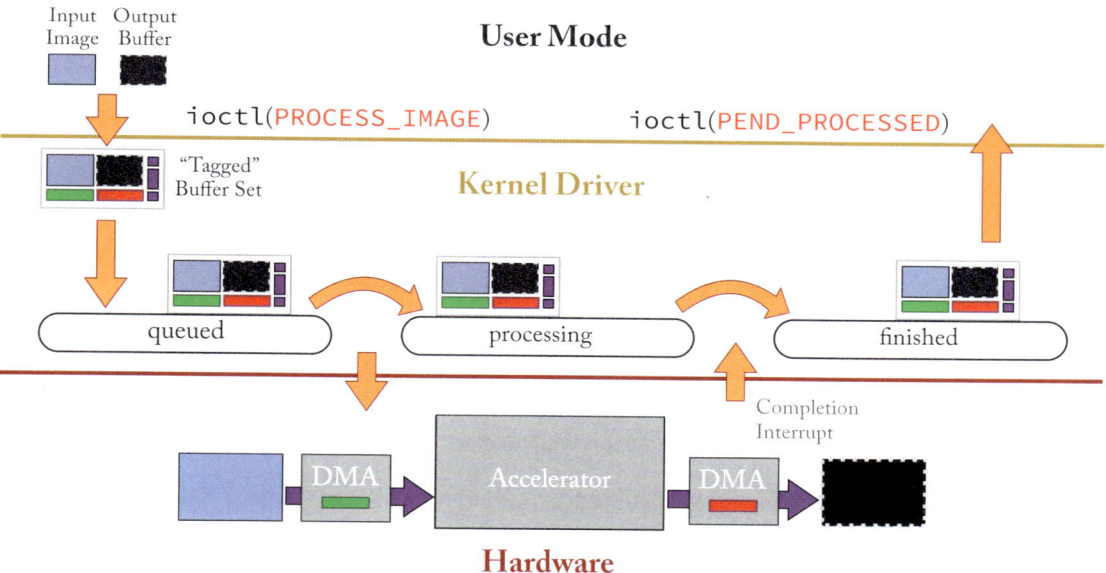

Figure 6.2: Internal operation of the driver. PROCESS_IMAGE places a buffer set onto the queue. Each time a completion interrupt fires, the driver starts the next buffer processing and puts the completed one on the output queue.

6.4.3 GENERATING SYSTEMS AND DRIVERS

Both Darkroom and Halide produce Verilog designs of accelerators, but to use them, we need a complete hardware system that provides interfaces to memory and the processor. To build this system, we apply a flexible template to create a wrapper around the accelerator. Every FPGA design from Halide and Darkroom is different, but because of the domain restriction, their interfaces all follow the same pattern: there are a number of input streams, a number of output streams, and some register space for tap values. We use the Mako template engine to create a Tcl script for Xilinx's Vivado toolchain. The Tcl script instantiates the accelerator, connects DMA engines for each input and output stream, wires up the interrupt lines, and runs the project build process to produce a bitstream.

We additionally automate the creation of design-specific boot collateral (first-stage bootloader and device tree) and package it up with the rest of the software stack (u-boot, Linux kernel, and root file system).

We can do the same thing for the software drivers, which must be configured with the correct numbers of streams, addresses of the DMA engines, interrupt lines, and image sizes. The driver generator reads an accelerator configuration from a YAML file and again uses Mako to parameterize the driver following a template. This includes the internal buffer structures that track all of the tap values and I/O buffers, and the body of the PROCESS_IMAGE function. The result is C code for a kernel driver, which is compiled to produce the binary kernel module.

6.4.4 GENERATING THE WHOLE STACK WITH HALIDE

Up to this point we've described the operation of the driver entirely independently of the image processing DSL, and in fact, the same basic infrastructure was used for both Darkroom and Halide. However, Halide offers an additional opportunity due to its generality and the flexibility of its compiler infrastructure.

Because we are generating the hardware and driver in the context of a Halide function, we can go beyond merely generating the driver. The developer specified exactly where the hardware/software boundary should be, using the Halide schedule, and the compiler knows exactly what the driver interface is (i.e., the dotted box in Figure 5.1b). Therefore, the Halide compiler can automatically insert calls to the kernel driver at that boundary.

Where the CPU version of the Halide code would compute the result, the heterogeneous version simply calls the kernel driver to delegate the work to the hardware, as shown in Figure 6.3. From the perspective of user code, the call to the Halide function is the same except for needing to pass in a handle for the hardware device. The only noticeable difference is that the heterogeneous version runs much faster.

We overlap the execution of CPU cores and accelerator using Linux threads. After the output image has been tiled into smaller blocks, the processing pipeline of the small block is wrapped by a loop iterating over the tiles. A user can schedule the loop to run in parallel using

Halide's `parallel` primitive, as shown in line 16 of Figure 5.1a. After this, a typical program implementation looks like:

```
parallel foreach tile_index:
  retrieve pinned_buffers
  compute values in pinned_buffers...
  task_id = launch_accelerator(pinned_buffers)
  wait_sync(task_id)
  consume values in pinned_buffers...
  release pinned_buffer
```

During execution, a thread pool is created for launching the workers that run the loop body. Some threads use CPU cores to compute values in the buffer, launch the accelerator task, and then quickly get blocked in the `wait_sync` calls. While these threads sleep and the accelerator is running, other active threads can use the CPU cores.

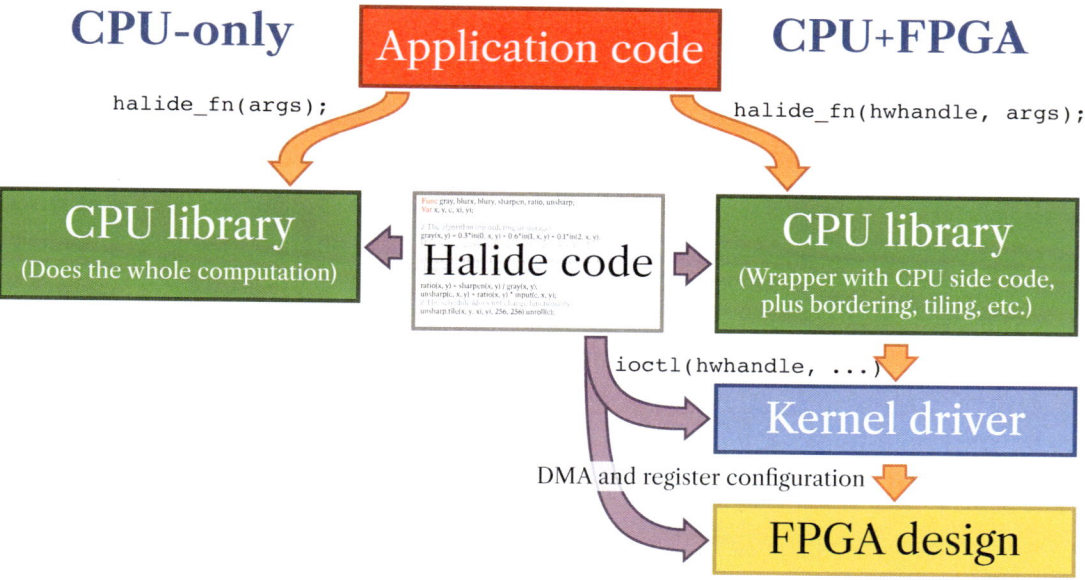

Figure 6.3: Call stack from userspace down to the hardware.

6.4.5 HETEROGENEOUS SYSTEM PERFORMANCE

System performance can be greatly improved if the CPU and the accelerator run concurrently while processing a heterogeneous pipeline. Figure 6.4 illustrates the effectiveness of our multi-threading technique described above, overlapping CPU and accelerator workloads for **stereo**.

stereo is a particularly ripe application for this, as the optimal partitioning of the pipeline results in roughly equal runtimes for the CPU and FPGA parts of the algorithm.

Figure 6.4 plots the program runtime using the same accelerator to process different size images with and without multiple threads. For each launch, the accelerator processes a 600×400 image tile in 7.63 ms. The software program breaks the image into such tiles, prepares an input image tile for the accelerator including padding the boundary of the original image, launches and waits for the accelerator, then repeats for the next tile.

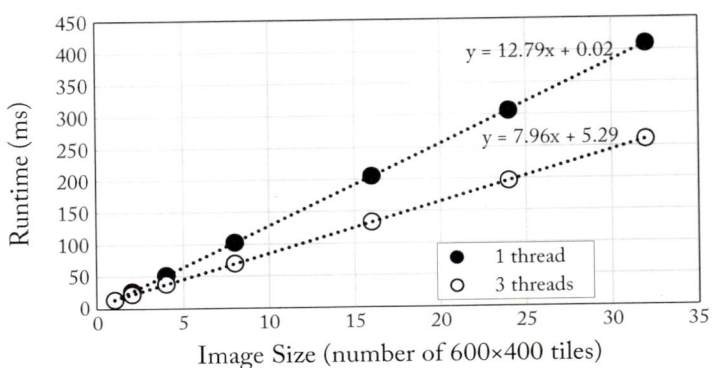

Figure 6.4: Runtime of **stereo** with different sizes of input image. All runs use a hardware accelerator that processes a 600 × 400 image tile in 7.93 ms. Our multithreading technique helps overlap CPU and accelerator execution time so that per-tile processing cost plus CPU overhead becomes just 7.96 ms instead of 12.76 ms.

When running the program sequentially with one thread, each tile takes 12.8 ms, indicating a CPU workload of about 5 ms/tile. However, using 3 threads to process 3 tiles concurrently, the per-tile processing cost reduces to 7.96 ms, as most of the CPU workload now overlaps with accelerator execution. With this overlap, the system is bounded by its slowest part, which in this case is the accelerator portion of the workload. When the CPU workload dominates, we may not see such nice behavior, as we find out below.

As described earlier, the accelerators are cache coherent and share the L2 cache with the CPU cores, thanks to the Zynq's ACP. If the intermediate blocks of memory used for passing data between CPU kernels and accelerators fit in the L2, the number of DRAM accesses will decrease, and the memory access latency will improve as well. To this end, blocking the image via the `tile` primitive effectively reduces the size of these CPU-accelerator intermediates.

Table 6.1 summarizes the performance and energy cost of **gaussian** compiled for different block sizes and pixel rates. (Note that the choice of application—in this case **gaussian**—is arbitrary as whatever hardware computation does not affect the results and only the memory traffic on the interface matters.) We show the accelerator execution time for both Verilog simulation and real time as measured when running the accelerator repeatedly on the Zynq. Memory accesses start to miss in the L2 when the block size exceeds 48 KB, causing the DRAM dynamic energy to increase from zero to around 3.3 nJ/pixel (i.e., 34 pJ/bit-access).[2]

Table 6.1: Performance and energy cost measurements of the **Gaussian** accelerators for different block sizes and pixel rates running at 100 MHz. Larger blocks cause L2 cache misses to DRAM, increasing DRAM dynamic energy.

Rate (px/cy)	Read BW (MB/s)	Block Size KB	—HW Time—		DRAM Dynamic Energy (nJ/px)
			Simulated (μs)	Measured (μs)	
1	300	48	193	199	0.0
		192	710	721	2.5
		768	2,730	2,764	3.3
2	600	48	99	111	0.0
		192	360	510	2.3
		768	1,378	2,444	3.3

When we compare simulated to measured accelerator performance (HW Time columns in Table 6.1), we see that for low rate configurations, the measured runtime does not degrade when large block sizes cause L2 misses since DMA buffers can hide the latency. However, for high rate configurations, although the accelerators are twice as fast as the unit-rate designs in simulation, peak speed is only achieved for the 48 KB block configuration (i.e., when memory accesses all hit in the L2). Because of cache misses, the actual runtime is 77% longer than the simulated accelerator runtime for the much larger 768 KB block configuration.

However, breaking images into many small blocks introduces two problems of its own: First, more overlapping block boundaries causes more recomputation; second, the host CPU has to schedule more accelerator launches through the device driver interface. On the current platform, we observe scheduling overhead around 100~200 microseconds per launch in the device driver, mostly caused by context switches and synchronizations between the user thread and kernel background threads responsible for managing accelerator launch and completion queues. As a result of these overheads, for most of the evaluated applications, fine-grained blocking to fit in the L2 is not an efficient strategy overall. Instead, we chose to block the image in larger sizes (e.g., 480 × 640) for our full system evaluation.

With further engineering of our existing system, it would be possible to increase memory bandwidth for streaming large blocks. However, if we were to design a new platform, we would

---

[2] Computing an RGB pixel in gaussian requires six bytes of memory access. If the memory access misses in the L2, we assume it causes two DRAM accesses.

choose a faster CPU core (the Zynq's 667 MHz A9 was slow even at its introduction in 2012), a larger shared L2 cache, and a hardware launch engine that pulls accelerator tasks directly from a memory buffer without CPU intervention, as opposed to having a background thread pushing tasks to the accelerator.

CHAPTER 7

Conclusions and Future Directions

Computer architecture is moving toward specialization, and we desperately need new tools, techniques, and languages to enable productivity in this new era. We have argued that to move forward, it is not sufficient to build more sophisticated compilers to automatically parallelize and retarget code (although such tools are certainly useful and welcome!). Instead, it will be more effective to create new languages optimized for particular domains, and write compilers that leverage domain knowledge about particularly good design patterns to produce great designs. Specialized hardware engines exploit unique features of the application to be as efficient as possible; our programming languages ought to do this too.

We've shown through the examples of Darkroom and Halide that the restrictions imposed by a domain-specific language do in fact allow the compiler to produce high-performance designs for multiple hardware targets, and that the code is actually portable between targets with little to no modification. And we've shown that these languages are expressive enough to implement many real image-processing applications.

There is plenty that could be done to improve these systems and the process of compiling applications for heterogeneous systems more generally. One exciting area of work is on frameworks that can automatically partition algorithms to run on heterogeneous systems. The recently developed autoscheduling system for Halide [Mullapudi et al., 2016] can produce schedules for Halide code on CPU that rival schedules written by experts, and runs in a matter of seconds. Building a comparable model and optimizer for a mixed CPU/GPU/FPGA system is challenging, but not out of reach. Such a system would be extremely valuable. The nuances of CPU performance are already difficult to comprehend, and the complexity of heterogeneous systems and scale of the accelerator design space make it all but impossible to optimize by hand.

While we have focused on image processing in this book, there are other areas where domain-specific languages and specialized hardware have a role to play. A prominent example is deep learning, which has a handful of frameworks that function much like domain-specific languages, including Caffe, Torch, and TensorFlow. Specialized hardware for deep learning is becoming increasingly common, with prominent commercial examples such as Google's Tensor Processing Units and NVIDIA's "tensor cores" integrated into their Tesla GPUs.

Languages for building languages, such as Delite [Sujeeth et al., 2014] and Terra [DeVito et al., 2013], are another avenue for continued research. An obvious downside to domain-specific languages is that every domain needs one, which means a great many languages to create, and a great many languages for developers to learn. Tools that provide a common language and core for creating DSLs will hasten their uptake.

Pushing compute performance forward into tomorrow's parallel, specialized, and heterogeneous future will not be easy. But as focus shifts from keeping up with Moore's law to making better use of what it has already given us, new ideas in compilers and design tools will keep the innovation going.

Bibliography

Andrew Adams, Eino-Ville Talvala, Sung Hee Park, David E. Jacobs, Boris Ajdin, Natasha Gelfand, Jennifer Dolson, Daniel Vaquero, Jongmin Baek, Marius Tico, Hendrik P. A. Lensch, Wojciech Matusik, Kari Pulli, Mark Horowitz, and Marc Levoy. The Frankencamera: An experimental platform for computational photography. *ACM Transactions on Graphics*, 29(4):1–12, July 2010. DOI: 10.1145/1778765.1778766. 63

Altera. Intel FPGA SDK for OpenCL. https://www.altera.com/products/design-software/embedded-software-developers/opencl/overview.html, 2016. 22

Aptina. Aptina MT9P111. http://www.onsemi.com/pub_link/Collateral/MT9P111-D.PDF, 2016. [Online; accessed December 30, 2016]. 48

Amazon AWS. EC2 F1 instances. https://aws.amazon.com/ec2/instance-types/f1/, 2017. 6

Jon Bentley. Programming pearls: Little languages. *Communications of the ACM*, 29(8):711–721, August 1986. DOI: 10.1145/6424.315691. 25

Jean-Yves Bouguet. Pyramidal implementation of the affine Lucas Kanade feature tracker. Technical report, Intel Corporation, 2001. 31

John S. Brunhaver. *Design and Optimization of a Stencil Engine*. Ph.D. thesis, Stanford University, 2015. 14, 37

Calin Caşcaval, Siddhartha Chatterjee, Hubertus Franke, Kevin J. Gildea, and Pratap Pattnaik. A taxonomy of accelerator architectures and their programming models. *IBM Journal of Research and Development*, 54(5):5:1–5:10, September 2010. DOI: 10.1147/JRD.2010.2059721. 69

John Canny. A computational approach to edge detection. *IEEE Transactions on Pattern Analysis and Machine Intelligence*, (6):679–698, 1986. DOI: 10.1016/b978-0-08-051581-6.50024-6. 29

Riccardo Cattaneo, Giuseppe Natale, Carlo Sicignano, Donatella Sciuto, and Marco Domenico Santambrogio. On how to accelerate iterative stencil loops: A scalable streaming-based approach. *ACM Transactions on Architecture and Code Optimization*, 12(4):53:1–53:26, December 2015. DOI: 10.1145/2842615. 22

Jiawen Chen, Sylvain Paris, and Frédo Durand. Real-time edge-aware image processing with the bilateral grid. *ACM Transactions on Graphics*, 26(3), July 2007. DOI: 10.1145/1276377.1276506. 31

Wenjie Chen, Zhibin Wang, Qin Wu, Jiuzhen Liang, and Zhilei Chai. Implementing dense optical flow computation on a heterogeneous FPGA SoC in C. *ACM Transactions on Architecture and Code Optimization*, 13(3):25:1–25:25, August 2016. DOI: 10.1145/2948976. 22

Philippe Coussy, Daniel D. Gajski, Michael Meredith, and Andres Takach. An introduction to high-level synthesis. *IEEE Design and Test of Computers*, 26(4):8–17, 2009. DOI: 10.1109/mdt.2009.69. 21

Andrew Danowitz, Kyle Kelley, James Mao, John P. Stevenson, and Mark Horowitz. CPU DB: Recording microprocessor history. *Communication of the ACM*, 55(4):55–63, April 2012. DOI: 10.1145/2133806.2133822. 4

Robert H. Dennard, Fritz H. Gaensslen, V. Leo Rideout, Ernest Bassous, and Andre R. LeBlanc. Design of ion-implanted MOSFET's with very small physical dimensions. *IEEE Journal of Solid-State Circuits*, 9(5):256–268, 1974. DOI: 10.1142/9789814503464_0086. 2

Zachary DeVito, James Hegarty, Alex Aiken, Pat Hanrahan, and Jan Vitek. Terra: A multistage language for high-performance computing. In *Proc. of the 34th ACM SIGPLAN Conference on Programming Language Design and Implementation*, pages 105–116, 2013. DOI: 10.1145/2491956.2462166. 42, 81

Jayanth Gummaraju and Mendel Rosenblum. Stream programming on general-purpose processors. In *Proc. of the 38th Annual IEEE/ACM International Symposium on Microarchitecture, (MICRO)*, pages 343–354, IEEE Computer Society, Washington, DC, 2005. DOI: 10.1109/MICRO.2005.32. 44

Chris Harris and Mike Stephens. A combined corner and edge detector. In *Proc. of the 4th Alvey Vision Conference*, pages 147–151, 1988. DOI: 10.5244/c.2.23. 29

James Hegarty, John Brunhaver, Zachary DeVito, Jonathan Ragan-Kelley, Noy Cohen, Steven Bell, Artem Vasilyev, Mark Horowitz, and Pat Hanrahan. Darkroom: Compiling high-level image processing code into hardware pipelines. *ACM Transactions on Graphics*, 33(4):144:1–11, July 2014. DOI: 10.1145/2601097.2601174. 33

International Business Strategies, IBS July 2017 monthly report: Design Activities and Strategic Implications, July 2017. www.ibs-inc.net 11

ITRS. International technology roadmap for semiconductors 2.0, 2015 edition. http://www.itrs2.net/itrs-reports.html, 2016. 1

Jonathan R. Kelley, Connelly Barnes, Andrew Adams, Sylvain Paris, Saman Amarasinghe, et al. Halide: A language and compiler for optimizing parallelism, locality, and recomputation in image processing pipelines. *SIGPLAN Notices*, 48(6):519–530, 2013. DOI: 10.1145/2499370.2462176. 20

Moein Khazraee, Lu Zhang, Luis Vega, and Michael Bedford Taylor. Moonwalk: NRE optimization in ASIC clouds. In *Proc. of the 22nd International Conference on Architectural Support for Programming Languages and Operating Systems, (ASPLOS)*, pages 511–526, ACM, New York, 2017. DOI: 10.1145/3037697.3037749. 10

Chris Lattner and Vikram Adve. LLVM: A Compilation Framework for Lifelong Program Analysis and Transformation. In *Proc. of the International Symposium on Code Generation and Optimization (CGO)*, Palo Alto, CA, March 2004. 46

Charles E. Leiserson and James B. Saxe. Retiming synchronous circuitry. *Algorithmica*, 6(1-6): 5–35, 1991. DOI: 10.1007/bf01759032. 42

lp_solve contributors. lp_solve library. http://lpsolve.sourceforge.net, 2010. 42

Bruce D. Lucas, Takeo Kanade, et al. An iterative image registration technique with an application to stereo vision. In *IJCAI*, volume 81, pages 674–679, 1981. 30

Dejan Marković and Robert W. Brodersen. *DSP Architecture Design Essentials*. Springer, 2012. 10

Richard Membarth and Oliver Reiche. Fork of HIPAcc generating code for Vivado HLS. https://github.com/hipacc/hipacc-vivado, 2016. 66

Richard Membarth, Oliver Reiche, Frank Hannig, Jürgen Teich, Mario Körner, and Wieland Eckert. HIPAcc: A domain-specific language and compiler for image processing. *IEEE Transactions on Parallel and Distributed Systems*, 27(1):210–224, 2016. DOI: 10.1109/tpds.2015.2394802. 68

Mentor Graphics. Catapult high-level synthesis. https://www.mentor.com/hls-lp/catapult-high-level-synthesis/, 2016. 22, 59

Gordon E. Moore. Cramming more components onto integrated circuits. *Electronics*, 38(8), April 1965. DOI: 10.1109/jproc.1998.658762. 2

Thierry Moreau, Mark Wyse, Jacob Nelson, Adrian Sampson, Hadi Esmaeilzadeh, Luis Ceze, and Mark Oskin. SNNAP: Approximate computing on programmable SoCs via neural acceleration. In *IEEE 21st International Symposium on High Performance Computer Architecture (HPCA)*, pages 603–614, San Francisco, CA, February 2015. DOI: 10.1109/HPCA.2015.7056066. 22

Ravi Teja Mullapudi, Andrew Adams, Dillon Sharlet, Jonathan Ragan-Kelley, and Kayvon Fatahalian. Automatically scheduling Halide image processing pipelines. *ACM Transactions on Graphics*, 35(4):83:1–83:11, July 2016. DOI: 10.1145/2897824.2925952. 81

Norman P. Jouppi et al. In-datacenter performance analysis of a tensor processing unit. In *Proc. of the 44th Annual International Symposium on Computer Architecture, (ISCA)*, pages 1–12, ACM, New York, 2017. DOI: 10.1145/3079856.3080246. 6

M. Akif Özkan, Oliver Reiche, Frank Hannig, and Jürgen Teich. FPGA-based accelerator design from a domain-specific language. In *26th International Conference on Field Programmable Logic and Applications (FPL)*, pages 1–9, IEEE, Lausanne, Switzerland, August 2016. DOI: 10.1109/FPL.2016.7577357. 67

Sylvain Paris and Frédo Durand. A fast approximation of the bilateral filter using a signal processing approach. *International Journal of Computer Vision*, 81(1):24–52, 2009. DOI: 10.1007/s11263-007-0110-8. 31

Jonathan Ragan-Kelley, Connelly Barnes, Andrew Adams, Sylvain Paris, Frédo Durand, and Saman Amarasinghe. Halide: A language and compiler for optimizing parallelism, locality, and recomputation in image processing pipelines. In *Proc. of the 34th ACM SIGPLAN Conference on Programming Language Design and Implementation, (PLDI)*, pages 519–530, New York, 2013. DOI: 10.1145/2491956.2462176. 58

William Hadley Richardson. Bayesian-based iterative method of image restoration. *JOSA*, 62 (1):55–59, 1972. DOI: 10.1364/josa.62.000055. 30

Ofer Shacham, Megan Wachs, Andrew Danowitz, Sameh Galal, John Brunhaver, Wajahat Qadeer, Sabarish Sankaranarayanan, Artem Vassiliev, Stephen Richardson, and Mark Horowitz. Avoiding game over: Bringing design to the next level. In *Proc. of the 49th Annual Design Automation Conference (DAC)*, pages 623–629, 2012. DOI: 10.1145/2228360.2228472. 44

Arvind K. Sujeeth, Kevin J. Brown, Hyoukjoong Lee, Tiark Rompf, Hassan Chafi, Martin Odersky, and Kunle Olukotun. Delite: A compiler architecture for performance-oriented embedded domain-specific languages. *ACM Transactions on Embedded Computer Systems*, 13 (4s):134:1–134:25, April 2014. DOI: 10.1145/2584665. 81

Maxeler Acceleration Technology. MaxCompiler White Paper. https://www.maxeler.com/media/documents/MaxelerWhitePaperMaxCompiler.pdf, 2011. 22

Michael Wolfe. Beyond induction variables. In *Proc. of the ACM SIGPLAN Conference on Programming Language Design and Implementation, (PLDI)*, pages 162–174, New York, 1992. DOI: 10.1145/143095.143131. 60

Xilinx. Vivado high-level synthesis. `http://www.xilinx.com/products/design-tools/vivado/integration/esl-design.html`, 2016. 22, 59, 66

Chen Zhang, Peng Li, Guangyu Sun, Yijin Guan, Bingjun Xiao, and Jason Cong. Optimizing FPGA-based accelerator design for deep convolutional neural networks. In *Proc. of the ACM/SIGDA International Symposium on Field-Programmable Gate Arrays, (FPGA)*, pages 161–170, New York, 2015. DOI: 10.1145/2684746.2689060. 22

Zhiru Zhang, Yiping Fan, Wei Jiang, Guoling Han, Changqi Yang, and Jason Cong. *AutoPilot: A Platform-Based ESL Synthesis System*, pages 99–112. Springer Netherlands, Dordrecht, 2008. DOI: 10.1007/978-1-4020-8588-8_6. 22

Authors' Biographies

STEVEN BELL

Steven Bell is a Ph.D. candidate at Stanford University, where he's building camera platforms as a vehicle to explore the challenge of rapidly creating high-performance hardware/software systems. As part of his Ph.D. work, he has developed imaging algorithms, written kernel drivers, and wrangled FPGA tools. His interests include image processing and computational photography, embedded software and systems, and teaching these topics to others. He received a B.S. in computer engineering from Oklahoma Christian University in 2011.

JING PU

Jing Pu received a B.S. in microelectronics from Peking University and an M.S. and Ph.D. in electrical engineering from Stanford University. He is currently working on Pixel Visual Core and Halide at Google. During his Ph.D., he worked on programming CPU/FPGA heterogeneous systems from the Halide image processing language. His research interests include domain-specific architectures and compilers, computer vision, and deep learning.

JAMES HEGARTY

James Hegarty is a research scientist whose work looks at developing new hardware design methodologies to enable rapid hardware-software co-design of low power image processing and computer vision systems. James' recent work involved the development of the Darkroom and Rigel image processing hardware languages. Previously, James worked on graphics systems, where he helped develop a novel real-time GPU rendering system that reduced shading load by 9x on micropolygon workloads. James holds a B.S., M.S., and Ph.D. in computer science from Stanford University.

MARK HOROWITZ

Mark Horowitz is the Yahoo! Founders Professor at Stanford University and was chair of the Electrical Engineering Department from 2008–2012. He co-founded Rambus, Inc. in 1990 and is a fellow of the IEEE and the ACM and a member of the National Academy of Engineering and the American Academy of Arts and Science. Dr. Horowitz's research interests are quite broad and span applying EE and CS analysis methods to problems in molecular biology to creating new design methodologies for analog and digital VLSI circuits.